# Montessori Math Workbook



## PRIMARY BOOK 3

A HANDS-ON APPROACH TO EARLY MATHEMATICS

by Katie Key
montessoriforhomeschoolers.com

ISBN: 978-1-7353980-6-8

Montessori Math Workbook - Primary Book 3

Graphics Used:
-Bead Clipart is from Bee Creative Cliparts.

montessoriforhomeschoolers.com
raisingkingdomwarriors.com

# Policy Page

By purchasing and/or downloading this workbook, you are agreeing to the following terms and conditions:

**Family License:**

1. This Workbook is to be used only by the immediate family of the purchaser. You may not distribute by giving or selling this workbook in any format to any person outside of your immediate family.
2. You may print as many copies for your immediate family as you would like.
3. Please do not use this workbook for multiple children in a classroom or co-op. If you need a set of physical Workbooks for your class or a license to use the digital version of this Workbook for your class, please email me directly at montessoriforhomeschoolers@gmail.com for bulk pricing.
4. Please do not sell or re-use any or all parts of this workbook for profit or otherwise.

**Classroom License:**

1. By purchasing a digital classroom license, you have permission to make copies for a class of up to 30 students for **one year**.
2. You may print as many copies for your **single classroom** to be used by up to 30 students in a single class for one year.
3. You may distribute the copies digitally, but please include this page with your distribution, and let your students' parents know that they cannot use or distribute this product in any way for any purposes other than completing the assignment for your class.
4. If you need a set of physical Workbooks for your class or to upgrade to a license to use the digital version of this Workbook for your class, please email me directly at montessoriforhomeschoolers@gmail.com for bulk pricing.
5. Please do not sell or re-use any or all parts of this workbook for profit or otherwise.

I trust that you will adhere to these policies, as it helps support a family like yours! I have worked very hard to create amazing resources for my family and yours, and I want to thank you for doing your part to support my small business!

# Dear Parent/Teacher,

Thank you for purchasing the *Montessori Math Workbook - Primary Book 3*! Whether you are continuing along your Montessori math journey or you are finding this "new" method of doing math and diving in now, welcome! I am honored to help facilitate a small part of your education journey.

This is the *third* Workbook in an expanding collection of Montessori Math Workbooks! This workbook is written for the child who has mastered the Decimal System with the cards and Golden Bead material, as well as the intial counting and formation of numbers exercises. Ideally, the child should work with the Golden Bead material *before* attempting the work in this workbook. If you need the Golden Bead Work, check out the *Montessori Math Workbook - Primary Book 2*.

We will start our passage to abstraction by moving from completely concrete numerals (i.e. the Golden Beads) to a *representation* of numerals with the Stamp Game.

Students will also start the *memorization* strand of the Montessori Math sequence by repeated practice with addition and subtraction facts using boards and charts, all found within the pages of this workbook.

This is not a "Definitive Guide to Montessori Math," rather it is an on-paper solution for the parent or teacher looking for an affordable, easy way to incorporate Montessori math at home or in the classroom.

You can use this workbook as a **supplement** to traditional Montessori math work or as a **stand-alone** math curriculum.

I welcome and encourage feedback! I am human, so if there are errors, please do not hesitate to let me know. If you see a way that the workbook can be improved, I welcome those comments as well. We also can *all* benefit from encouragement, so I humbly ask for you to write me your praises so I can be encouraged, and so I can share them on my sales page if you give me permission.

Enjoy this journey with your child! May the Lord bless you and keep you, shine His face upon you, and give you peace.

In Christ,

Katie Key
montessoriforhomeschoolers@gmail.com

# Math Objectives Covered

In Primary Book 3, your child will be learn:

- addition facts 1 through 18 via the Montessori Addition Strip Board and Charts, as well as the exercises with the Bead Houses,
- subtraction facts 1 through 18 via the Montessori Subtraction Strip Board and the Bead Houses,
- addition, subtraction, multiplication, and division of 2-4 digit numbers using the Montessori Stamp Game, with emphasis on exchanging,
- whole, half, and quarter fractions,
- how to tell time to the hour, half hour, quarter hour, and minute,
- how to read a calendar and write the date,

PLUS a looping review of previously covered math concepts, such as:

- writing numbers and number words
- number rods and operations with number rods
- even and odd
- greater than and less than
- the short bead stair and operations with the short bead stair
- teen number recognition and formation
- counting by tens
- counting 1-100 using the hundred chain and the hundred board
- squaring chains and skip counting
- the decimal system using the Golden Bead Material
- formation of numerals using the Golden Bead Material and decimal cards
- addition, subtraction, multiplication, and division using the Golden Bead Material
- addition strip board problems
- addition working chart problems
- subtraction strip board problems
- bead houses

By working through this Math Workbook, a child will move along toward further abstraction by working with hands-on manipulatives and visuals, as well as inadvertently memorize math facts of addition and subtraction. Handwriting is more critical at this stage, but students are still welcome to use scribes (the teacher/parent) when needed.

Celebrate your child's accomplishment in completing this workbook by presenting the special letter found on the very last page. You will give your child specific praise for his or her hard work, and you can record the date completed for your records.

# The Hands-On, Self-Correcting Components

Montessori education for the primary and elementary ages is rooted hands-on work and scientifically developed for the child to hone age-appropriate skills as independently as possible.

You will find most work in here to be "hands-on" and require the physical manipulation of components. I have provided printable options in this workbook that you can cut out and laminate or re-inforce with poster board, recycled cardboard, or even balsa wood. I am not as creative as most, so I would just laminate the materials and store them alongside the Workbook in an accompanying binder with plastic page protectors to hold the pieces and charts.

The child will learn to self-check her work using the provided charts and techniques presented in the workbook. There are some story problem exercises that suggest the child check her work with her "guide." You may want to verify that the child sets up the problem correctly.

For the Stamp Game exercises, you will find that the Control of Error when using the boxes is helpful. If a child adds 9+6 in the units column, only one digit can be placed in the green units box, so your child will be visually reminded to exchange ten units for one ten. I have also included half-page Keys for each of the Stamp Game lessons that you can use at your discretion in the Appendix.

# How to Progress through this Workbook

**Option #1: Cover to Cover**
**108 Days**
Simply open up the Workbook and complete the lessons in order! There are 80 main content lessons in this Workbook. If you factor in 2 days for each of the longer Stamp Game lessons, it brings the total main lesson count to 83.

When you get to the Looping Review Exercises, you may want to do one *page* a day instead of one colored box each day. If you do one page a day of the Looping Review Exercises, they will be completed in 24 school days. (Note: If you choose this route, you will only want to fill out one date, one start time, and one end time per page.)

Spend one day celebrating the completion of the Workbook for a total of 108 work days!

To spread the *Montessori Math Workbook 3* over one school year (36 weeks,) finish 3 lessons **or** 3 Looping Review Pages each week.

**Option #2: New Material Plus Integrated Review (*Suggested*)**
**144 Days**
This option is great for integrated review of past material. It also gives you some built-in "light" math days, along with more repetitive practice of telling time and reading a calendar. Remember, you can schedule the material in the Workbook however works best for your family, and this is only a suggestion!

There are 83 main lessons (see explanation in Option #1), 60 Looping Review Exercises, and 1 day of celebration at the completion of the workbook, which gives a total of 144 days of work. Working four days each week on math allows the *Montessori Math Workbook 3* to be completed in 36 weeks - or one school year!

Units 10 and 11 should be completed *before* doing the Looping Review Exercises, so do those every other day until those units are completed, and *then* move on to the Looping Review Exercises every other day. See the Suggested Schedule below.

Schedule Suggestion:
**Day 1:** New Lesson
**Day 2:** Unit 10 or 11 Lesson or Looping Review Exercise
**Day 3:** New Lesson
**Day 4:** Unit 10 or 11 Lesson or Looping Review Exercise

**Option #3: Allow Your Child to Set the Pace**
**? Days**

Children at the Primary age can truly be in charge of their work choices. Keep modeling, encouraging, and presenting lessons regularly, but follow your child's natural pace!

Note: If you have an elementary age child working through this Workbook, you may want to stick to Option #1 or #2, with the exception being a child who wants to work more quickly. If your child is still struggling to complete lessons happily, try setting a timer and encouraging full attention for that agreed upon time. You will work then in time increments rather than numbered-lesson increments.

# Support Website

My goal in writing all of the directions and visual instructions for this workbook was to make it as "stand-alone" as possible for you! That being said, I know that some of us learn better (and more quickly) when we see someone do the exercise in front of us. So, I created some instructional videos to supplement the on-paper directions!

Go to: https://montessoriforhomeschoolers.com/pages/math-book-3 for video instructions on some of the lessons, as well as any other supporting documents that may be released post-print.

Note: I will also post links to the physical Montessori materials that coincide with this workbook, should you choose to incorporate them into your learning environment. The materials are *not required* for a successful Montessori math experience using this Workbook! Some families just prefer to use them in addition to the Workbook.

# TABLE OF CONTENTS

# Addition Strip Board Instructions

In the Appendix you will find the **Addition Strip Board (A-1)**, as well as the **red and blue number strips (A-3)** and the **Addition Control Chart 1 (A-5).** Remove these from the workbook and laminate them for durability. Carefully cut out the number strips. These will be used for many activities. These specific instructions for setting up the Addition Strip Board should be followed for Lessons 1.1-1.5.

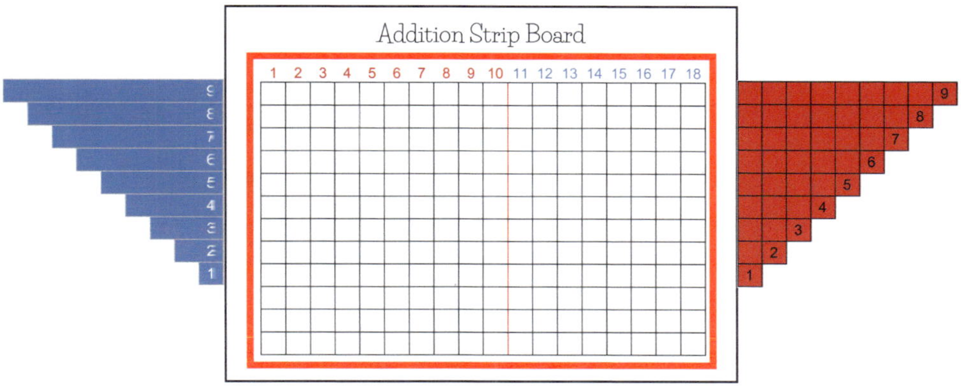


**1** Set up the Addition Strip Board with "angel wings," the blue strips on the left for the constant and the red strips on the right.

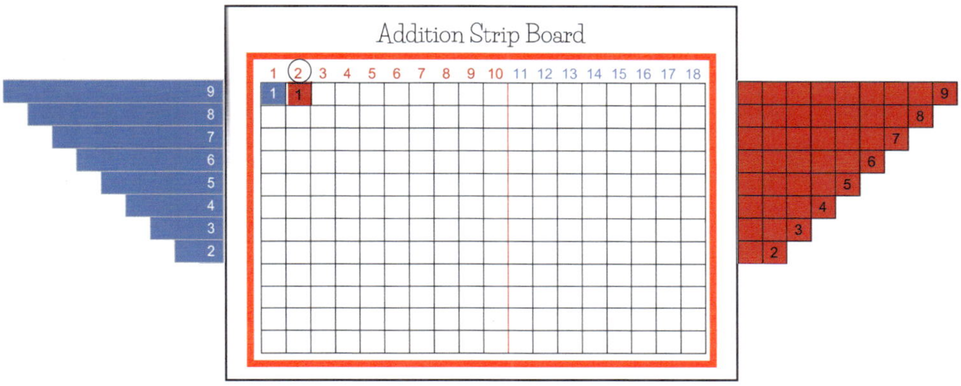


**2** Read the problem. Place the blue strip representing the first addend on the top row of the board. Then place red strip representing the second addend directly to the right of the blue strip.

The **sum** is found by reading the number at the top of the strip board where the red strip ends (circled).

Record the sum.

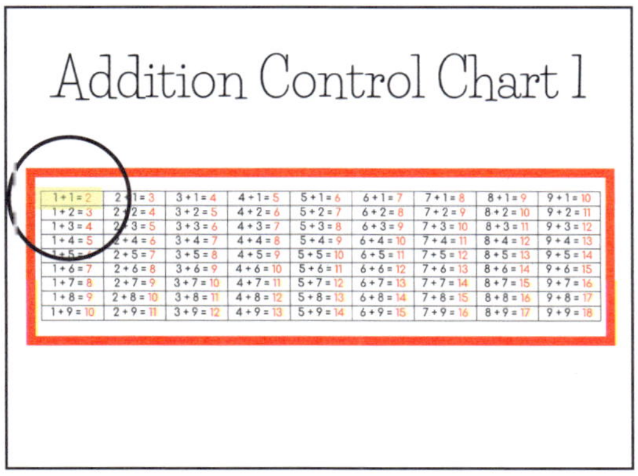


**3** Check your answer using the **Addition Control Chart 1.**

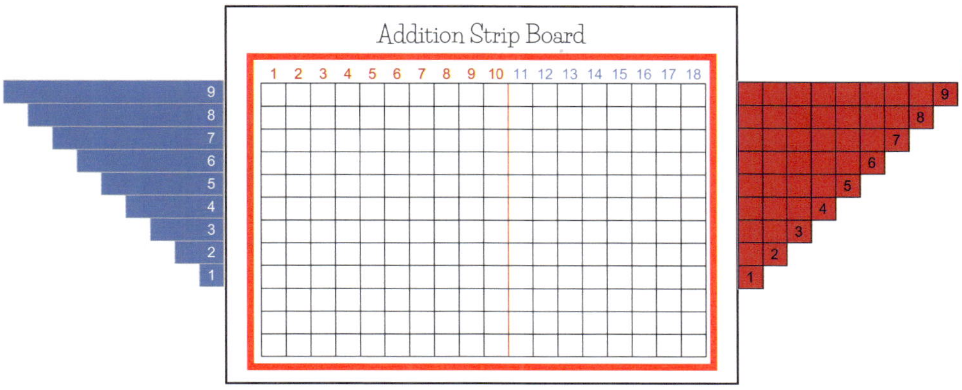


**4** Reset the board for the next problem. Repeat steps 1-3 for the next problem.

## Addition Strip Board: 1

**addition table of**

1 **one** 1

**Directions:**
**Part 1:** Follow the instructions in the Unit 1 Introduction to complete the Addition Table using the **Addition Strip Board** and the **Addition Control Chart 1.**
**Part 2:** Use the addition strips to the right of each problem below to redo and verify your work. **Color the first addend (constant) in blue** and the **second addend in red**. The number that the red strip ends directly underneath is the sum. Circle the sum.
**Part 3:** Check that the sum from the physical strip board and the colored-in addition strips match. If they do not, check the work both ways to discover the error.

| Problem | Addition strip | The sums match |
|---|---|---|
| 1 + 1 = 2 | 1 2 3 4 5 6 7 8 9 10 11 12 13 14 15 16 17 18 | ✓ |
| 1 + 2 = ____ | 1 2 3 4 5 6 7 8 9 10 11 12 13 14 15 16 17 18 | ○ |
| 1 + 3 = ____ | 1 2 3 4 5 6 7 8 9 10 11 12 13 14 15 16 17 18 | ○ |
| 1 + 4 = ____ | 1 2 3 4 5 6 7 8 9 10 11 12 13 14 15 16 17 18 | ○ |
| 1 + 5 = ____ | 1 2 3 4 5 6 7 8 9 10 11 12 13 14 15 16 17 18 | ○ |
| 1 + 6 = ____ | 1 2 3 4 5 6 7 8 9 10 11 12 13 14 15 16 17 18 | ○ |
| 1 + 7 = ____ | 1 2 3 4 5 6 7 8 9 10 11 12 13 14 15 16 17 18 | ○ |
| 1 + 8 = ____ | 1 2 3 4 5 6 7 8 9 10 11 12 13 14 15 16 17 18 | ○ |
| 1 + 9 = ____ | 1 2 3 4 5 6 7 8 9 10 11 12 13 14 15 16 17 18 | ○ |

## Addition Strip Board: 2

**addition table of**

2 **two** 2

| Problem | Addition strip | The sums match. |
|---|---|---|
| 2 + 1 = ____ | 1 2 3 4 5 6 7 8 9 10 11 12 13 14 15 16 17 18 | ○ |
| 2 + 2 = ____ | 1 2 3 4 5 6 7 8 9 10 11 12 13 14 15 16 17 18 | ○ |
| 2 + 3 = ____ | 1 2 3 4 5 6 7 8 9 10 11 12 13 14 15 16 17 18 | ○ |
| 2 + 4 = ____ | 1 2 3 4 5 6 7 8 9 10 11 12 13 14 15 16 17 18 | ○ |
| 2 + 5 = ____ | 1 2 3 4 5 6 7 8 9 10 11 12 13 14 15 16 17 18 | ○ |
| 2 + 6 = ____ | 1 2 3 4 5 6 7 8 9 10 11 12 13 14 15 16 17 18 | ○ |
| 2 + 7 = ____ | 1 2 3 4 5 6 7 8 9 10 11 12 13 14 15 16 17 18 | ○ |
| 2 + 8 = ____ | 1 2 3 4 5 6 7 8 9 10 11 12 13 14 15 16 17 18 | ○ |
| 2 + 9 = ____ | 1 2 3 4 5 6 7 8 9 10 11 12 13 14 15 16 17 18 | ○ |

# 1.2 | Addition Strip Board Tables of 3 & 4

## Addition Strip Board: 3

**Directions:**
**Part 1:** Follow the instructions in the Unit 1 Introduction to complete the Addition Table using the **Addition Strip Board** and the **Addition Control Chart 1.**
**Part 2:** Use the addition strips to the right of each problem below to redo and verify your work. **Color the first addend (constant) in blue** and the **second addend in red**. The number that the red strip ends directly underneath is the sum. Circle the sum.
**Part 3:** Check that the sum from the physical strip board and the colored-in addition strips match. If they do not, check the work both ways to discover the error.

addition table
of
3 three 3

| Problem | Strip | The sums match. |
|---|---|---|
| 3 + 1 = ____ | 1 2 3 4 5 6 7 8 9 10 11 12 13 14 15 16 17 18 | ◯ |
| 3 + 2 = ____ | 1 2 3 4 5 6 7 8 9 10 11 12 13 14 15 16 17 18 | ◯ |
| 3 + 3 = ____ | 1 2 3 4 5 6 7 8 9 10 11 12 13 14 15 16 17 18 | ◯ |
| 3 + 4 = ____ | 1 2 3 4 5 6 7 8 9 10 11 12 13 14 15 16 17 18 | ◯ |
| 3 + 5 = ____ | 1 2 3 4 5 6 7 8 9 10 11 12 13 14 15 16 17 18 | ◯ |
| 3 + 6 = ____ | 1 2 3 4 5 6 7 8 9 10 11 12 13 14 15 16 17 18 | ◯ |
| 3 + 7 = ____ | 1 2 3 4 5 6 7 8 9 10 11 12 13 14 15 16 17 18 | ◯ |
| 3 + 8 = ____ | 1 2 3 4 5 6 7 8 9 10 11 12 13 14 15 16 17 18 | ◯ |
| 3 + 9 = ____ | 1 2 3 4 5 6 7 8 9 10 11 12 13 14 15 16 17 18 | ◯ |

## Addition Strip Board: 4

addition table
of
4 four 4

| Problem | Strip | The sums match. |
|---|---|---|
| 4 + 1 = ____ | 1 2 3 4 5 6 7 8 9 10 11 12 13 14 15 16 17 18 | ◯ |
| 4 + 2 = ____ | 1 2 3 4 5 6 7 8 9 10 11 12 13 14 15 16 17 18 | ◯ |
| 4 + 3 = ____ | 1 2 3 4 5 6 7 8 9 10 11 12 13 14 15 16 17 18 | ◯ |
| 4 + 4 = ____ | 1 2 3 4 5 6 7 8 9 10 11 12 13 14 15 16 17 18 | ◯ |
| 4 + 5 = ____ | 1 2 3 4 5 6 7 8 9 10 11 12 13 14 15 16 17 18 | ◯ |
| 4 + 6 = ____ | 1 2 3 4 5 6 7 8 9 10 11 12 13 14 15 16 17 18 | ◯ |
| 4 + 7 = ____ | 1 2 3 4 5 6 7 8 9 10 11 12 13 14 15 16 17 18 | ◯ |
| 4 + 8 = ____ | 1 2 3 4 5 6 7 8 9 10 11 12 13 14 15 16 17 18 | ◯ |
| 4 + 9 = ____ | 1 2 3 4 5 6 7 8 9 10 11 12 13 14 15 16 17 18 | ◯ |

## Addition Strip Board: 5

**Directions:**
**Part 1:** Follow the instructions in the Unit 1 Introduction to complete the Addition Table using the **Addition Strip Board** and the **Addition Control Chart 1.**
**Part 2:** Use the addition strips to the right of each problem below to redo and verify your work. **Color the first addend (constant) in blue** and the **second addend in red**. The number that the red strip ends directly underneath is the sum. Circle the sum.
**Part 3:** Check that the sum from the physical strip board and the colored-in addition strips match. If they do not, check the work both ways to discover the error.

addition table
of
5 five 5

5 + 1 = ______
5 + 2 = ______
5 + 3 = ______
5 + 4 = ______
5 + 5 = ______
5 + 6 = ______
5 + 7 = ______
5 + 8 = ______
5 + 9 = ______

| 1 | 2 | 3 | 4 | 5 | 6 | 7 | 8 | 9 | 10 | 11 | 12 | 13 | 14 | 15 | 16 | 17 | 18 | The sums match. |
|---|---|---|---|---|---|---|---|---|---|---|---|---|---|---|---|---|---|---|
| 1 | 2 | 3 | 4 | 5 | 6 | 7 | 8 | 9 | 10 | 11 | 12 | 13 | 14 | 15 | 16 | 17 | 18 | ○ |
| 1 | 2 | 3 | 4 | 5 | 6 | 7 | 8 | 9 | 10 | 11 | 12 | 13 | 14 | 15 | 16 | 17 | 18 | ○ |
| 1 | 2 | 3 | 4 | 5 | 6 | 7 | 8 | 9 | 10 | 11 | 12 | 13 | 14 | 15 | 16 | 17 | 18 | ○ |
| 1 | 2 | 3 | 4 | 5 | 6 | 7 | 8 | 9 | 10 | 11 | 12 | 13 | 14 | 15 | 16 | 17 | 18 | ○ |
| 1 | 2 | 3 | 4 | 5 | 6 | 7 | 8 | 9 | 10 | 11 | 12 | 13 | 14 | 15 | 16 | 17 | 18 | ○ |
| 1 | 2 | 3 | 4 | 5 | 6 | 7 | 8 | 9 | 10 | 11 | 12 | 13 | 14 | 15 | 16 | 17 | 18 | ○ |
| 1 | 2 | 3 | 4 | 5 | 6 | 7 | 8 | 9 | 10 | 11 | 12 | 13 | 14 | 15 | 16 | 17 | 18 | ○ |
| 1 | 2 | 3 | 4 | 5 | 6 | 7 | 8 | 9 | 10 | 11 | 12 | 13 | 14 | 15 | 16 | 17 | 18 | ○ |
| 1 | 2 | 3 | 4 | 5 | 6 | 7 | 8 | 9 | 10 | 11 | 12 | 13 | 14 | 15 | 16 | 17 | 18 | ○ |

## Addition Strip Board: 6

addition table
of
6 six 6

6 + 1 = ______
6 + 2 = ______
6 + 3 = ______
6 + 4 = ______
6 + 5 = ______
6 + 6 = ______
6 + 7 = ______
6 + 8 = ______
6 + 9 = ______

| 1 | 2 | 3 | 4 | 5 | 6 | 7 | 8 | 9 | 10 | 11 | 12 | 13 | 14 | 15 | 16 | 17 | 18 | The sums match. |
|---|---|---|---|---|---|---|---|---|---|---|---|---|---|---|---|---|---|---|
| 1 | 2 | 3 | 4 | 5 | 6 | 7 | 8 | 9 | 10 | 11 | 12 | 13 | 14 | 15 | 16 | 17 | 18 | ○ |
| 1 | 2 | 3 | 4 | 5 | 6 | 7 | 8 | 9 | 10 | 11 | 12 | 13 | 14 | 15 | 16 | 17 | 18 | ○ |
| 1 | 2 | 3 | 4 | 5 | 6 | 7 | 8 | 9 | 10 | 11 | 12 | 13 | 14 | 15 | 16 | 17 | 18 | ○ |
| 1 | 2 | 3 | 4 | 5 | 6 | 7 | 8 | 9 | 10 | 11 | 12 | 13 | 14 | 15 | 16 | 17 | 18 | ○ |
| 1 | 2 | 3 | 4 | 5 | 6 | 7 | 8 | 9 | 10 | 11 | 12 | 13 | 14 | 15 | 16 | 17 | 18 | ○ |
| 1 | 2 | 3 | 4 | 5 | 6 | 7 | 8 | 9 | 10 | 11 | 12 | 13 | 14 | 15 | 16 | 17 | 18 | ○ |
| 1 | 2 | 3 | 4 | 5 | 6 | 7 | 8 | 9 | 10 | 11 | 12 | 13 | 14 | 15 | 16 | 17 | 18 | ○ |
| 1 | 2 | 3 | 4 | 5 | 6 | 7 | 8 | 9 | 10 | 11 | 12 | 13 | 14 | 15 | 16 | 17 | 18 | ○ |
| 1 | 2 | 3 | 4 | 5 | 6 | 7 | 8 | 9 | 10 | 11 | 12 | 13 | 14 | 15 | 16 | 17 | 18 | ○ |

## Addition Strip Board: 7

**Directions:**
**Part 1:** Follow the instructions in the Unit 1 Introduction to complete the Addition Table using the **Addition Strip Board** and the **Addition Control Chart 1.**
**Part 2:** Use the addition strips to the right of each problem below to redo and verify your work. **Color the first addend (constant) in blue** and the **second addend in red**. The number that the red strip ends directly underneath is the sum. Circle the sum.
**Part 3:** Check that the sum from the physical strip board and the colored-in addition strips match. If they do not, check the work both ways to discover the error.

addition table
of
7 seven 7

| Problem | Addition strip | The sums match. |
|---|---|---|
| 7 + 1 = ______ | 1 2 3 4 5 6 7 8 9 10 11 12 13 14 15 16 17 18 | ○ |
| 7 + 2 = ______ | 1 2 3 4 5 6 7 8 9 10 11 12 13 14 15 16 17 18 | ○ |
| 7 - 3 = ______ | 1 2 3 4 5 6 7 8 9 10 11 12 13 14 15 16 17 18 | ○ |
| 7 + 4 = ______ | 1 2 3 4 5 6 7 8 9 10 11 12 13 14 15 16 17 18 | ○ |
| 7 + 5 = ______ | 1 2 3 4 5 6 7 8 9 10 11 12 13 14 15 16 17 18 | ○ |
| 7 + 6 = ______ | 1 2 3 4 5 6 7 8 9 10 11 12 13 14 15 16 17 18 | ○ |
| 7 + 7 = ______ | 1 2 3 4 5 6 7 8 9 10 11 12 13 14 15 16 17 18 | ○ |
| 7 + 8 = ______ | 1 2 3 4 5 6 7 8 9 10 11 12 13 14 15 16 17 18 | ○ |
| 7 + 9 = ______ | 1 2 3 4 5 6 7 8 9 10 11 12 13 14 15 16 17 18 | ○ |

## Addition Strip Board: 8

addition table
of
8 eight 8

| Problem | Addition strip | The sums match. |
|---|---|---|
| 8 + 1 = ______ | 1 2 3 4 5 6 7 8 9 10 11 12 13 14 15 16 17 18 | ○ |
| 8 + 2 = ______ | 1 2 3 4 5 6 7 8 9 10 11 12 13 14 15 16 17 18 | ○ |
| 8 + 3 = ______ | 1 2 3 4 5 6 7 8 9 10 11 12 13 14 15 16 17 18 | ○ |
| 8 + 4 = ______ | 1 2 3 4 5 6 7 8 9 10 11 12 13 14 15 16 17 18 | ○ |
| 8 + 5 = ______ | 1 2 3 4 5 6 7 8 9 10 11 12 13 14 15 16 17 18 | ○ |
| 8 + 6 = ______ | 1 2 3 4 5 6 7 8 9 10 11 12 13 14 15 16 17 18 | ○ |
| 8 + 7 = ______ | 1 2 3 4 5 6 7 8 9 10 11 12 13 14 15 16 17 18 | ○ |
| 8 + 8 = ______ | 1 2 3 4 5 6 7 8 9 10 11 12 13 14 15 16 17 18 | ○ |
| 8 + 9 = ______ | 1 2 3 4 5 6 7 8 9 10 11 12 13 14 15 16 17 18 | ○ |

# 1.5 | Addition Strip Board Table of 9

## Addition Strip Board: 9

**Directions:**
**Part 1:** Follow the instructions in the Unit 1 Introduction to complete the Addition Table using the **Addition Strip Board** and the **Addition Control Chart 1.**
**Part 2:** Use the addition strips to the right of each problem below to redo and verify your work. **Color the first addend (constant) in blue** and the **second addend in red**. The number that the red strip ends directly underneath is the sum. Circle the sum.
**Part 3:** Check that the sum from the physical strip board and the colored-in addition strips match. If they do not, check the work both ways to discover the error.

addition table
of
9 nine 9

| Problem | 1 | 2 | 3 | 4 | 5 | 6 | 7 | 8 | 9 | 10 | 11 | 12 | 13 | 14 | 15 | 16 | 17 | 18 | The sums match. |
|---|---|---|---|---|---|---|---|---|---|---|---|---|---|---|---|---|---|---|---|
| 9 + 1 = ______ | | | | | | | | | | | | | | | | | | | ○ |
| 9 + 2 = ______ | | | | | | | | | | | | | | | | | | | ○ |
| 9 + 3 = ______ | | | | | | | | | | | | | | | | | | | ○ |
| 9 + 4 = ______ | | | | | | | | | | | | | | | | | | | ○ |
| 9 + 5 = ______ | | | | | | | | | | | | | | | | | | | ○ |
| 9 + 6 = ______ | | | | | | | | | | | | | | | | | | | ○ |
| 9 + 7 = ______ | | | | | | | | | | | | | | | | | | | ○ |
| 9 + 8 = ______ | | | | | | | | | | | | | | | | | | | ○ |
| 9 + 9 = ______ | | | | | | | | | | | | | | | | | | | ○ |

# Addition Strip Board: Ways to Make 10

**Directions:**

1. Invite your student to cut out the red number strips on the next page and arrange them in order to the right of the workbook (like the right "angel wing" as in previous exercises.)
2. Explain that you are going to **make ten** by adding a red strip to each blue strip on the addition strip board below.
3. Demonstrate how to find the proper strip so when placed end to end, the blue and red strips stop at the red line after the 10.
4. Check to make sure the equation is correct using the **Addition Control Chart 1.**
5. Record the equation on the grid paper on the following page using blue, red, and black pencils. (See example.)
6. Continue until all rows add to ten.

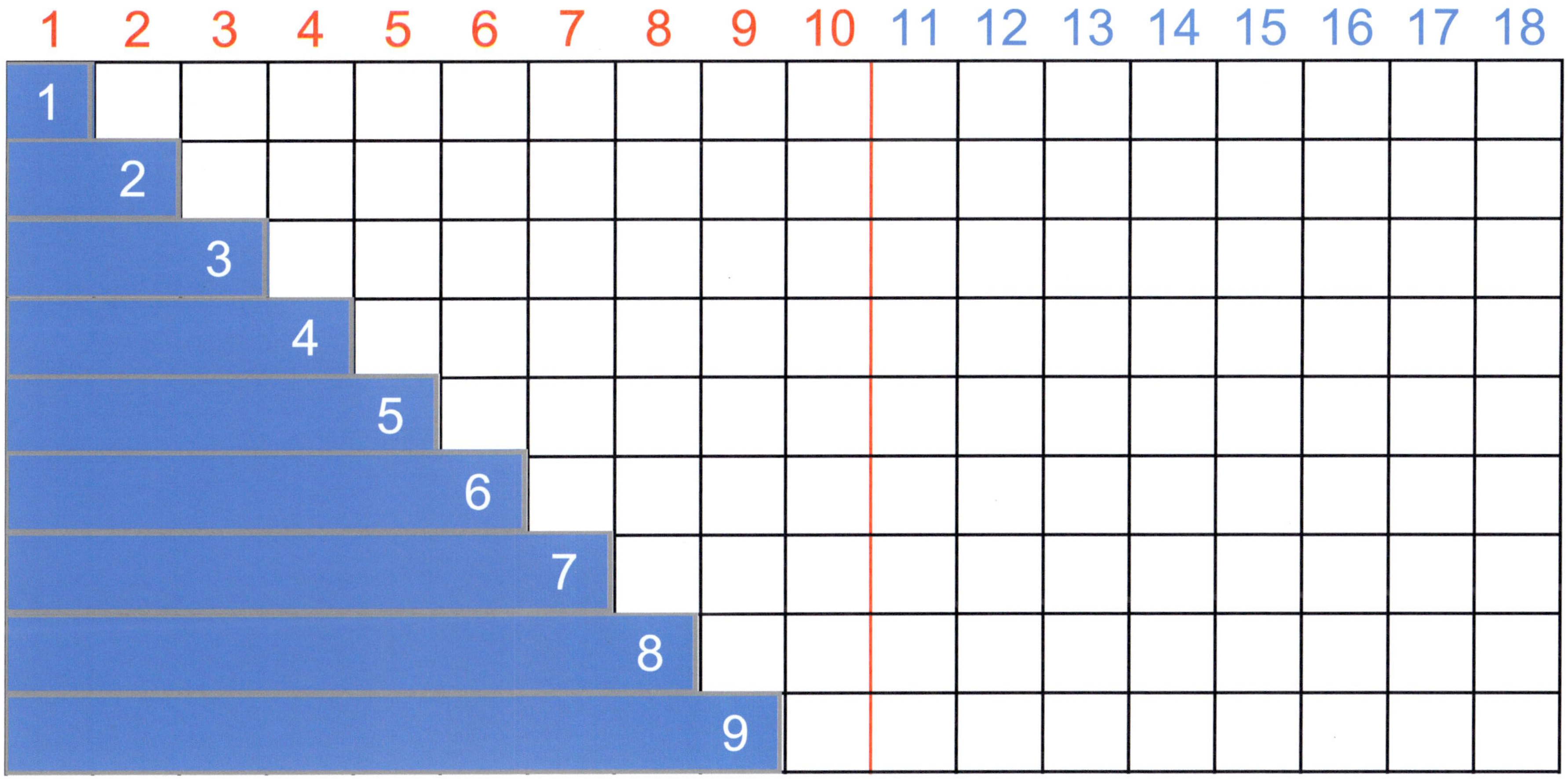

**Note:** See directions on the previous page. Record only one digit per square to keep your work neat and orderly.

| | | | | | |
|---|---|---|---|---|---|
| 1 | + | 9 | = | 1 | 0 |
| | | | | | |
| | | | | | |
| | | | | | |
| | | | | | |

| | | | | | |
|---|---|---|---|---|---|
| | | | | | |
| | | | | | |
| | | | | | |
| | | | | | |
| | | | | | |

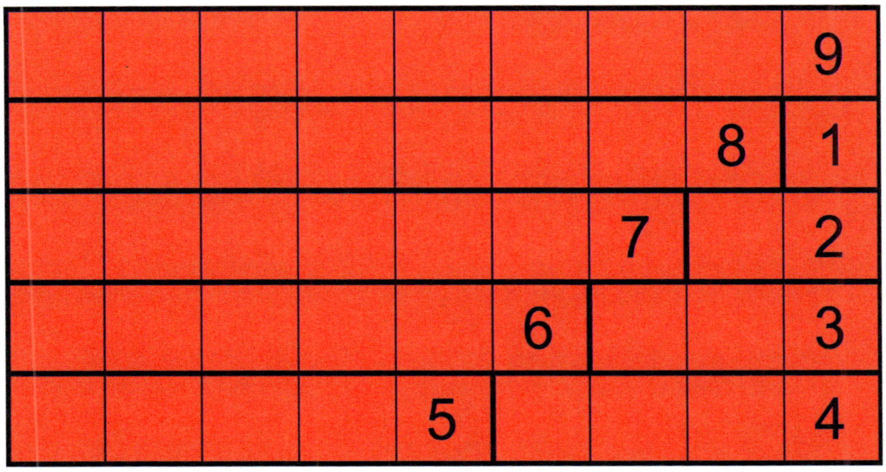

# 1.7 | Addition Strip Board: Deleting Duplicate Equations (Part 1)

In the previous lesson, we found out all of the ways to make ten using the addition strips. In this lesson, we are going to look for **duplicates**. This is a hands-on exercise that shows children there are fewer equations to memorize in math than it may at first seem.

**Directions:**

1. Repeat the exercise from the last lesson, but this time use your physical copy of the **Addition Strip Board** that you have laminated.
   a. Start by setting out your blue and red number strips as "angel wings," with blue on the left and red on the right. (See the layout below for a visual of the proper setup.)
   b. Next, find all the combinations of ten, starting with the blue "1" and the red "9." Go in order of the blue strips (1, 2, 3, 4, etc.) until all the strips are placed just as in the previous lesson.
2. Say, "Since 1 + 9 = 10 is the same as 9 + 1 = 10, we can take away this last equation (9 + 1), because it is a **duplicate**." Remove the blue 9 strip and the red 1 strip and replace them to the sides of the board in their proper angel wing positions.
3. **Now draw a line through the 9 + 1 = 10 equation on your grid paper in the previous lesson.**
4. Next, point out the 8+2, "Look, here is another one. 8 + 2 = 10 is the same as 2 + 8 = 10, so it is a duplicate. We can remove the 8 + 2 since we have the same equation up here." (Point to 2 + 8 when you say "here.")
5. **Now draw a line through the 8 + 2 = 10 equation on your grid paper in the previous lesson.**
6. Repeat with 7 + 3 and 6 + 4 (removing the ones lower on the chart, so the blue 7 and blue 6.) Make sure to remind your child to cross out the equations on the grid paper.
7. When you get to 5 + 5, show that there is no other 5 + 5 duplicate, so you do not need to remove anything.

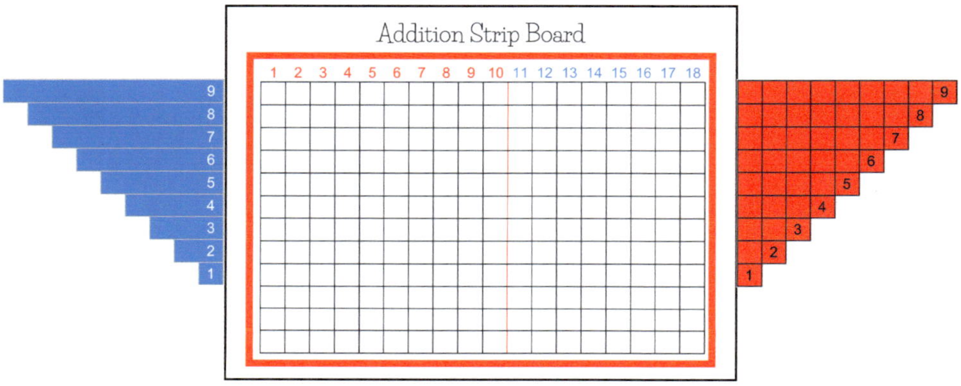

# 1.8 | Addition Strip Board: Deleting Duplicate Equations (Part 2)

**Directions:**

In the last lesson, we found duplicate equations for numbers that add to ten. We crossed out the duplicates, because we don't want to clutter up our brains with "extra" equations!

We are now going to do the same thing to our **Addition Control Chart 1**!

**Step 1:**

1. Show your child the laminated **Addition Control Chart 1.** Tell your child it is the same as the one on the next page of the workbook that we are going to be writing on for this activity.
2. Starting with the first column, circle the first equation. If you find a duplicate in the first row, cross it out! (Note: You will only find one equation for 1 + 1 = 2.)
3. Now circle the second equation in the first column (1 + 2 = 3). Look for a duplicate equation in the first row. If you find a duplicate in the first row, cross it out!
4. Continue going down the first *column* and circling one equation at a time then crossing out the corresponding duplicate in the first *row*.
5. Continue circling and crossing out with the second *column* and second *row*, "keeping" the equations only in the vertical columns by circling them and "celeting" duplicate equations in the horizontal rows by crossing them out. .

**Circle and cross out like this:**

| | | | | | | | | |
|---|---|---|---|---|---|---|---|---|
| 1 + 1 = 2 | ~~2 + 1 = 3~~ | ~~3 + 1 = 4~~ | ~~4 + 1 = 5~~ | 5 + 1 = 6 | 6 + 1 = 7 | 7 + 1 = 8 | 8 + 1 = 9 | 9 + 1 = 10 |
| 1 + 2 = 3 | 2 + 2 = 4 | 3 + 2 = 5 | 4 + 2 = 6 | 5 + 2 = 7 | 6 + 2 = 8 | 7 + 2 = 9 | 8 + 2 = 10 | 9 + 2 = 11 |
| 1 + 3 = 4 | 2 + 3 = 5 | 3 + 3 = 6 | 4 + 3 = 7 | 5 + 3 = 8 | 6 + 3 = 9 | 7 + 3 = 10 | 8 + 3 = 11 | 9 + 3 = 12 |
| 1 + 4 = 5 | 2 + 4 = 6 | 3 + 4 = 7 | 4 + 4 = 8 | 5 + 4 = 9 | 6 + 4 = 10 | 7 + 4 = 11 | 8 + 4 = 12 | 9 + 4 = 13 |
| 1 + 5 = 6 | 2 + 5 = 7 | 3 + 5 = 8 | 4 + 5 = 9 | 5 + 5 = 10 | 6 + 5 = 11 | 7 + 5 = 12 | 8 + 5 = 13 | 9 + 5 = 14 |
| 1 + 6 = 7 | 2 + 6 = 8 | 3 + 6 = 9 | 4 + 6 = 10 | 5 + 6 = 11 | 6 + 6 = 12 | 7 + 6 = 13 | 8 + 6 = 14 | 9 + 6 = 15 |
| 1 + 7 = 8 | 2 + 7 = 9 | 3 + 7 = 10 | 4 + 7 = 11 | 5 + 7 = 12 | 6 + 7 = 13 | 7 + 7 = 14 | 8 + 7 = 15 | 9 + 7 = 16 |
| 1 + 8 = 9 | 2 + 8 = 10 | 3 + 8 = 11 | 4 + 8 = 12 | 5 + 8 = 13 | 6 + 8 = 14 | 7 + 8 = 15 | 8 + 8 = 16 | 9 + 8 = 17 |
| 1 + 9 = 10 | 2 + 9 = 11 | 3 + 9 = 12 | 4 + 9 = 13 | 5 + 9 = 14 | 6 + 9 = 15 | 7 + 9 = 16 | 8 + 9 = 17 | 9 + 9 = 18 |

**When finished, you should be left with the following equations:**

| | | | | | | | | |
|---|---|---|---|---|---|---|---|---|
| 1 + 1 = 2 | | | | | | | | |
| 1 + 2 = 3 | 2 + 2 = 4 | | | | | | | |
| 1 + 3 = 4 | 2 + 3 = 5 | 3 + 3 = 6 | | | | | | |
| 1 + 4 = 5 | 2 + 4 = 6 | 3 + 4 = 7 | 4 + 4 = 8 | | | | | |
| 1 + 5 = 6 | 2 + 5 = 7 | 3 + 5 = 8 | 4 + 5 = 9 | 5 + 5 = 10 | | | | |
| 1 + 6 = 7 | 2 + 6 = 8 | 3 + 6 = 9 | 4 + 6 = 10 | 5 + 6 = 11 | 6 + 6 = 12 | | | |
| 1 + 7 = 8 | 2 + 7 = 9 | 3 + 7 = 10 | 4 + 7 = 11 | 5 + 7 = 12 | 6 + 7 = 13 | 7 + 7 = 14 | | |
| 1 + 8 = 9 | 2 + 8 = 10 | 3 + 8 = 11 | 4 + 8 = 12 | 5 + 8 = 13 | 6 + 8 = 14 | 7 + 8 = 15 | 8 + 8 = 16 | |
| 1 + 9 = 10 | 2 + 9 = 11 | 3 + 9 = 12 | 4 + 9 = 13 | 5 + 9 = 14 | 6 + 9 = 15 | 7 + 9 = 16 | 8 + 9 = 17 | 9 + 9 = 18 |

**Step 2:**

1. Compare your "edited' **Addition Control Chart 1** with the **Addition Control Chart 2 (found on page A-7)**. Notice how the **Addition Control Chart 2** has no duplicate equations! Tell your student that there aren't actually 81 equations to memorize, but there are only 45! (Your student may want to count the number of equations on the Addition Control Chart 2 to verify the number of equations.)
2. You can also note that the **Addition Control Chart 2** is laid out differently than the edited **Addition Control Chart 1**. In the **Addition Control Chart 2**, the equations with the same sum are placed in the same row.

# Addition Control Chart 1

*Student will cross out equations on this sheet. (See instructions on previous page.)*

| | | | | | | | | |
|---|---|---|---|---|---|---|---|---|
| 1 + 1 = 2 | 2 + 1 = 3 | 3 + 1 = 4 | 4 + 1 = 5 | 5 + 1 = 6 | 6 + 1 = 7 | 7 + 1 = 8 | 8 + 1 = 9 | 9 + 1 = 10 |
| 1 + 2 = 3 | 2 + 2 = 4 | 3 + 2 = 5 | 4 + 2 = 6 | 5 + 2 = 7 | 6 + 2 = 8 | 7 + 2 = 9 | 8 + 2 = 10 | 9 + 2 = 11 |
| 1 + 3 = 4 | 2 + 3 = 5 | 3 + 3 = 6 | 4 + 3 = 7 | 5 + 3 = 8 | 6 + 3 = 9 | 7 + 3 = 10 | 8 + 3 = 11 | 9 + 3 = 12 |
| 1 + 4 = 5 | 2 + 4 = 6 | 3 + 4 = 7 | 4 + 4 = 8 | 5 + 4 = 9 | 6 + 4 = 10 | 7 + 4 = 11 | 8 + 4 = 12 | 9 + 4 = 13 |
| 1 + 5 = 6 | 2 + 5 = 7 | 3 + 5 = 8 | 4 + 5 = 9 | 5 + 5 = 10 | 6 + 5 = 11 | 7 + 5 = 12 | 8 + 5 = 13 | 9 + 5 = 14 |
| 1 + 6 = 7 | 2 + 6 = 8 | 3 + 6 = 9 | 4 + 6 = 10 | 5 + 6 = 11 | 6 + 6 = 12 | 7 + 6 = 13 | 8 + 6 = 14 | 9 + 6 = 15 |
| 1 + 7 = 8 | 2 + 7 = 9 | 3 + 7 = 10 | 4 + 7 = 11 | 5 + 7 = 12 | 6 + 7 = 13 | 7 + 7 = 14 | 8 + 7 = 15 | 9 + 7 = 16 |
| 1 + 8 = 9 | 2 + 8 = 10 | 3 + 8 = 11 | 4 + 8 = 12 | 5 + 8 = 13 | 6 + 8 = 14 | 7 + 8 = 15 | 8 + 8 = 16 | 9 + 8 = 17 |
| 1 + 9 = 10 | 2 + 9 = 11 | 3 + 9 = 12 | 4 + 9 = 13 | 5 + 9 = 14 | 6 + 9 = 15 | 7 + 9 = 16 | 8 + 9 = 17 | 9 + 9 = 18 |

# 1.9 | Addition Strip Board: Adding Zero

**Directions**:

1. Starting with the first row of the **Addition Strip Board** below, invite your student to color his own "number strips" for each equation given.
2. The student should then record the answer on the grid below the Addition Strip Board.
3. The student should continue with the next equation in the same manner, using the second row of the Addition Strip Board, and so on..
4. When your student finishes all of the equations in this lesson, ask, "What happens when we add zero to a number?" (The answer is, "The number stays the same.")

## Addition Strip Board

| 1 | 2 | 3 | 4 | 5 | 6 | 7 | 8 | 9 | 10 | 11 | 12 | 13 | 14 | 15 | 16 | 17 | 18 |
|---|---|---|---|---|---|---|---|---|---|---|---|---|---|---|---|---|---|
| | | | | | | | | | | | | | | | | | |
| | | | | | | | | | | | | | | | | | |
| | | | | | | | | | | | | | | | | | |
| | | | | | | | | | | | | | | | | | |
| | | | | | | | | | | | | | | | | | |
| | | | | | | | | | | | | | | | | | |
| | | | | | | | | | | | | | | | | | |
| | | | | | | | | | | | | | | | | | |
| | | | | | | | | | | | | | | | | | |
| | | | | | | | | | | | | | | | | | |
| | | | | | | | | | | | | | | | | | |
| | | | | | | | | | | | | | | | | | |

| | | | | | | | | | | | |
|---|---|---|---|---|---|---|---|---|---|---|---|
| 1 | + | 0 | = | | | 6 | + | 0 | = | | |
| 2 | + | 0 | = | | | 7 | + | 0 | = | | |
| 3 | + | 0 | = | | | 8 | + | 0 | = | | |
| 4 | + | 0 | = | | | 9 | + | 0 | = | | |
| 5 | + | 0 | = | | | | | | | | |

# 1.10 | Addition Strip Board: Doubles

**Directions**:

1. Cut the number strips apart and arrange them on your work space in Angel Wings.
2. Invite your student to discover the sum of the doubles of numbers, starting with 1 + 1 in the first row. Arrange the strips. Glue strips down. Find the sum. **Record the sum below the strip board.**
3. Repeat with 2 + 2 in the second row, 3 + 3 in the third row, etc., until 9 + 9 has been discovered.
4. Check the sums with the **Addition Control Chart 2** found on page **A-7.**

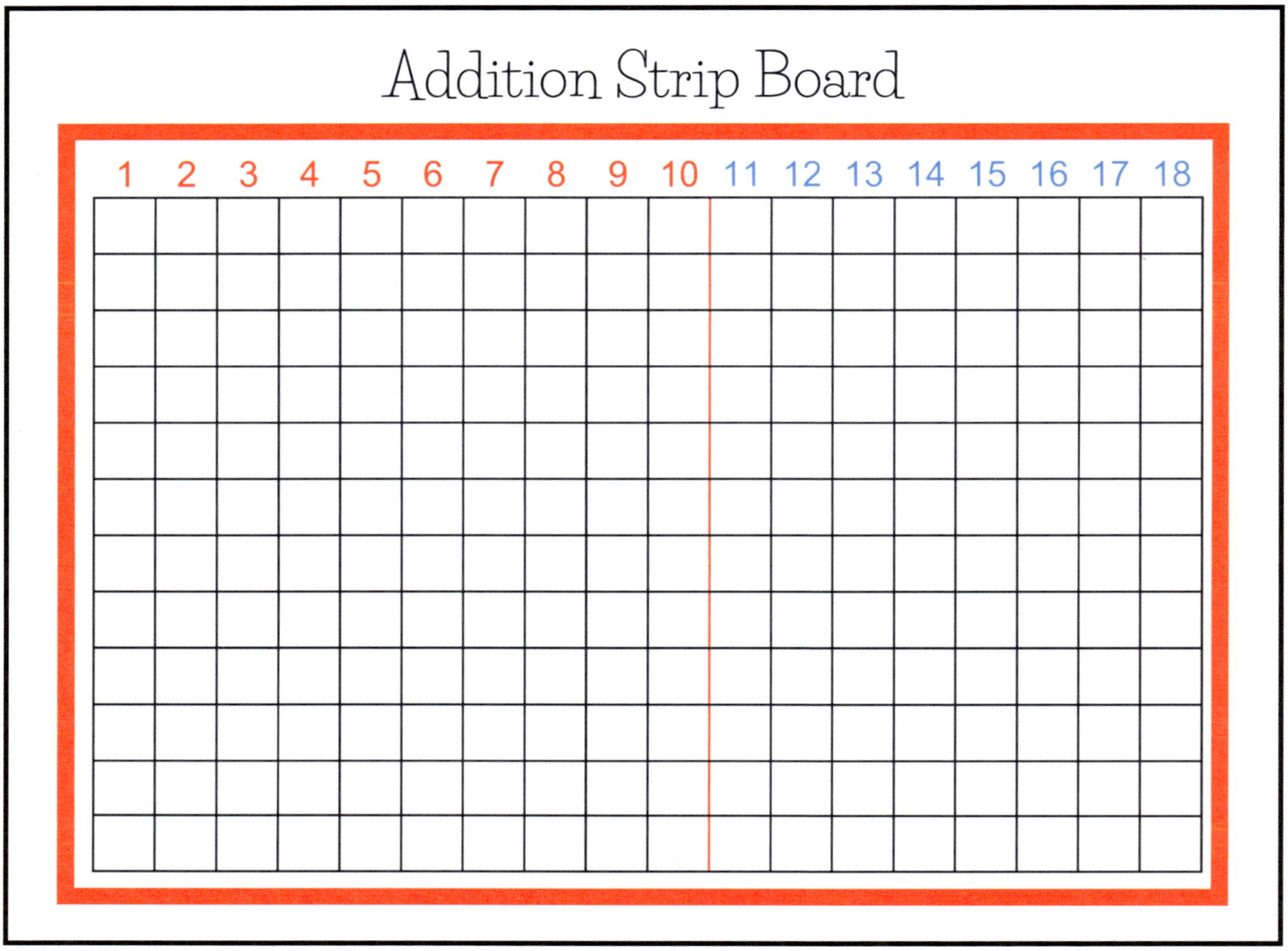


| | | | | | | | | | | | | |
|---|---|---|---|---|---|---|---|---|---|---|---|---|
| 1 | + | 1 | = | | | | 6 | + | 6 | = | | |
| 2 | + | 2 | = | | | | 7 | + | 7 | = | | |
| 3 | + | 3 | = | | | | 8 | + | 8 | = | | |
| 4 | + | 4 | = | | | | 9 | + | 9 | = | | |
| 5 | + | 5 | = | | | | | | | | | |

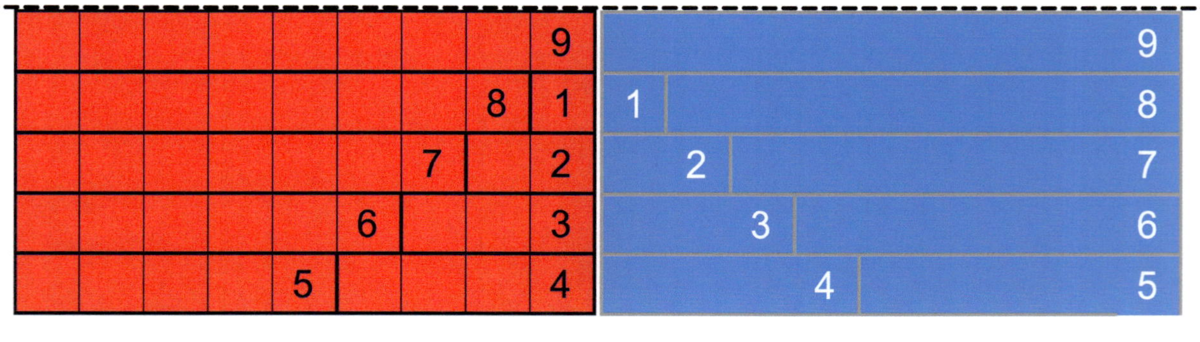

# 1.11 | Addition Strip Board: Story Problems 1

**Directions**:

1. Use the Addition Strip Board to solve these story problems.
2. Write the equation under each problem and the answer on the line.
3. Verify the answer using the **Addition Control Chart 2** and then color the picture if the answer is correct. (Redo the problem to correct it if the answer is incorrect *before* coloring the picture.)

Liana's mother is throwing Liana a birthday party. Liana is turning seven! She has already made 8 invitations for her friends. She has to make 6 more invitations. How many invitations will she make in all?

**Liana will make ________ invitations.**

Ella and Benji are playing cars. They made a huge track out of tape that winds around their city made of building blocks. Ella brought 9 cars, and Benji brought 5 cars. How many cars are they playing with all together?

**Ella and Beji are playing with _______ cars.**

Danny and John emptied out their lunch boxes and counted their items. Danny counted 7 items and John counted 4 items. How many items do the two boys have total?

**The boys have a total of ______ items in their lunches.**

# 1.12 | Addition Strip Board: Story Problems 2

**Directions**:

1. Use the Addition Strip Board to solve these story problems.
2. Write the equation under each problem and the answer on the line.
3. Verify the answer using the **Addition Control Chart 2** and then color the picture if the answer is correct. (Redo the problem to correct it if the answer is incorrect *before* coloring the picture.)

The girls were saving up their chore money to start a cake business. Jessie saved $8 and Sarah saved $3. How much money had the girls saved so far?

**The girls had saved ___________.**

The boys wanted to start a business, too. They started saving their coins in their penny banks. James had 9¢ and Ronald had 7¢. How much money had the boys saved altogether?

**The boys saved ____________.**

A group of friends went to play at the beach. They decided to have a shell-finding contest. They split up into two teams: one boy team and one girl team. At the end of two minutes, the girls had found 6 shells, and the boys had also found 6 shells. They tied! How many shells did they find in all?

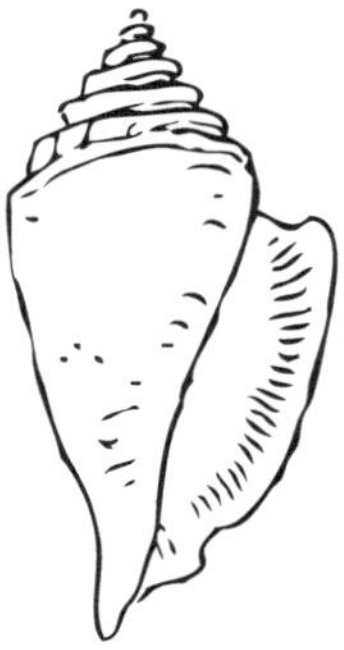

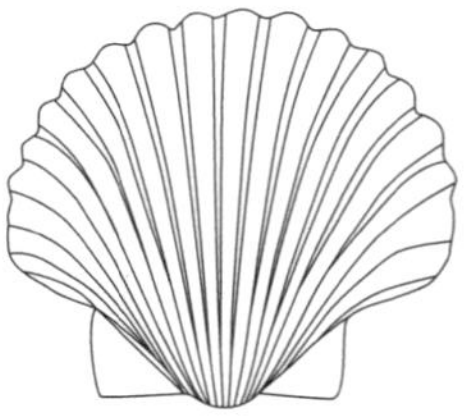

**The friends found ________ shells in all.**

# 2.1 | Addition Working Chart 3

**Directions**: We will practice sums again today using the **Addition Working Chart 3**, found on page **A-9**.

1. Cut up a few equations to work on today and place them in a small bowl or basket.
2. Allow your child to choose an equation then record it on the grid paper on the next page. If your student has a fact memorized and does not need the chart to answer the question, then she should write it in the "I know it!" column. If your child needs to use the chart, she should write it in the "I'm still practicing!" column.
3. Using the **Addition Working Chart 3**, find the sum then record it on the grid paper. (See step by step instructions in the below example.)

**Example: 3 + 7 =**

1. Place the index finger of your right hand on the blue numeral 3 (first addend) and the index finger of your left hand on the red numeral 7 (second addend).
2. Slowly move your left hand across the row and your right hand down the column until your two fingers meet in the same square. The square in which they meet is the sum.
3. Record 3 + 7 = 10 on grid paper in the appropriate column.*

Right index finger: 1st addend

Left index finger: 2nd addend

| 0 | 1 | 2 | 3 | 4 | 5 | 6 | 7 | 8 | 9 |
|---|---|---|---|---|---|---|---|---|---|
| 1 | 2 | 3 | 4 | 5 | 6 | 7 | 8 | 9 | 10 |
| 2 | 3 | 4 | 5 | 6 | 7 | 8 | 9 | 10 | 11 |
| 3 | 4 | 5 | 6 | 7 | 8 | 9 | 10 | 11 | 12 |
| 4 | 5 | 6 | 7 | 8 | 9 | 10 | 11 | 12 | 13 |
| 5 | 6 | 7 | 8 | 9 | 10 | 11 | 12 | 13 | 14 |
| 6 | 7 | 8 | 9 | 10 | 11 | 12 | 13 | 14 | 15 |
| 7 | 8 | 9 | 10 (sum) | 11 | 12 | 13 | 14 | 15 | 16 |
| 8 | 9 | 10 | 11 | 12 | 13 | 14 | 15 | 16 | 17 |
| 9 | 10 | 11 | 12 | 13 | 14 | 15 | 16 | 17 | 18 |

*write like this on the grid paper

| | | | | | |
|---|---|---|---|---|---|
| 3 | + | 7 | = | 1 | 0 |

| | | |
|---|---|---|
| 1 + 4 = | 5 + 1 = | 9 + 4 = |
| 3 + 9 = | 6 + 7 = | 7 + 2 = |
| 4 + 5 = | 7 + 8 = | 6 + 6 = |
| 3 + 7 = | 8 + 5 = | 2 + 1 = |
| 2 + 3 = | 9 + 1 = | 4 + 8 = |

# Addition Working Chart 3 - Grid Paper for Sums

I know it!

I'm still practicing.

# 2.2 | Addition Working Chart 3: Deleting Duplicates

**Materials:**

- ruler or straight edge
- highlighter
- Addition Working Chart 4 (page A-11)

**Directions:**

1. Using a straight edge, draw a light diagonal line from the top left corner to the bottom right corner of the Addition Working Chart 3 on the next page, as shown below in purple.
2. Tell your student, **"There are the same number of sums above the line as there are below the line. Today, we need to find the duplicates and delete them!"**
3. **"Let's start with the sum of 2."** Show, with your fingers, 1 + 1 = 2. **"1 + 1 = 2 is not a duplicate, so nothing can be deleted."**
4. **"Now let's do the sums of 3. 1 + 2 = 3."** (Put your index finger of one hand on the blue 1 (which is the first addend) and your index finger of your other hand on the red 2 (which is the second addend), and slide them to meet each other at the sum of 3.) **"Is there another combination of 3? [Yes.] Let's find the other sum of 3."** (Point to the 3 above the diagonal line.) **"2 + 1 = 3. Since we already have this equation, just with the addends flipped, we can delete this sum of 3!"** Cross out the 3 with a highlighter.
5. Continue, moving in order. You will find the sum(s) of 4 to 9 to the left of the diagonal first, then find the corresponding sum(s) of 4 to 9 on the right side of the diagonal and cross them out. When you are finished, the chart should have all of the purple sums (see below) crossed out.
6. Ask, **"How many sums are left?"** (Note the sums are in white. We do not count 1+ 0, 2+ 0, etc.. The answer is 45.)
7. Now show your child the Addition Working Chart 4. (Page **A-11**.) Ask, **"What do you notice that is the same about the Addition Working Chart 4 and the Addition Working Chart 3?"** (Allow child to answer, but make sure to point out that Chart 3 has the duplicate equations and Chart 4 has no duplicate equations.)

| 0 | 1 | 2 | 3 | 4 | 5 | 6 | 7 | 8 | 9 |
|---|---|---|---|---|---|---|---|---|---|
| 1 | 2 | 3 | 4 | 5 | 6 | 7 | 8 | 9 | 10 |
| 2 | 3 | 4 | 5 | 6 | 7 | 8 | 9 | 10 | 11 |
| 3 | 4 | 5 | 6 | 7 | 8 | 9 | 10 | 11 | 12 |
| 4 | 5 | 6 | 7 | 8 | 9 | 10 | 11 | 12 | 13 |
| 5 | 6 | 7 | 8 | 9 | 10 | 11 | 12 | 13 | 14 |
| 6 | 7 | 8 | 9 | 10 | 11 | 12 | 13 | 14 | 15 |
| 7 | 8 | 9 | 10 | 11 | 12 | 13 | 14 | 15 | 16 |
| 8 | 9 | 10 | 11 | 12 | 13 | 14 | 15 | 16 | 17 |
| 9 | 10 | 11 | 12 | 13 | 14 | 15 | 16 | 17 | 18 |

| 0 | 1 | 2 | 3 | 4 | 5 | 6 | 7 | 8 | 9 |
|---|---|---|---|---|---|---|---|---|---|
| 1 | 2 | 3 | 4 | 5 | 6 | 7 | 8 | 9 | 10 |
| 2 | 3 | 4 | 5 | 6 | 7 | 8 | 9 | 10 | 11 |
| 3 | 4 | 5 | 6 | 7 | 8 | 9 | 10 | 11 | 12 |
| 4 | 5 | 6 | 7 | 8 | 9 | 10 | 11 | 12 | 13 |
| 5 | 6 | 7 | 8 | 9 | 10 | 11 | 12 | 13 | 14 |
| 6 | 7 | 8 | 9 | 10 | 11 | 12 | 13 | 14 | 15 |
| 7 | 8 | 9 | 10 | 11 | 12 | 13 | 14 | 15 | 16 |
| 8 | 9 | 10 | 11 | 12 | 13 | 14 | 15 | 16 | 17 |
| 9 | 10 | 11 | 12 | 13 | 14 | 15 | 16 | 17 | 18 |

# 2.3 | Addition Working Chart 4

**Materials**

- **Addition Control Chart 1 or 2** (removed from Workbook)
- **Addition Working Chart 4** (removed from Workbook, page A-11)

**Directions**

1. Cut apart some or all of the equations and place them in a bowl or basket. (Note: You may want to cut only the number of equations you want your student to complete today.)
2. Demonstrate how to do the activity by first choosing a random equation from the basket yourself. Invite your student to record the equation on the grid paper on the next page in the left column.
3. Now, using the laminated **Addition Working Chart 4**, show your student how to find the sum of **<u>6 + 3 =</u>**.
   a. Place one index finger on the red numeral representing the first addend.
   b. Place the index finger of your opposite hand on the red numeral that represents the second addend.
   c. Slowly move each finger to the right until you get to the last square in each row.
   d. Now, "walk" your fingers toward each other at the same pace. You will either end up on a number (even) or in a blank space (odd). If you end up in a blank space, the answer will be the number one down and one to the left (Another way of describing the sum if in a blank spot is that it is one diagonal space down to the left.)
   e. Together with your student, check to make sure the sum is correct by comparing it to the **Addition Control Chart 1.**
   f. Invite your student to record the answer.
4. Invite your student to choose an equation from the basket and record it on the next page. (Note: You may want your student to label the columns, "I know it!" and "I'm still practicing." like the previous lesson. If he knows the answer, he can skip using the chart and just record the equation.)
5. Observe carefully as your student uses the chart the first few times.

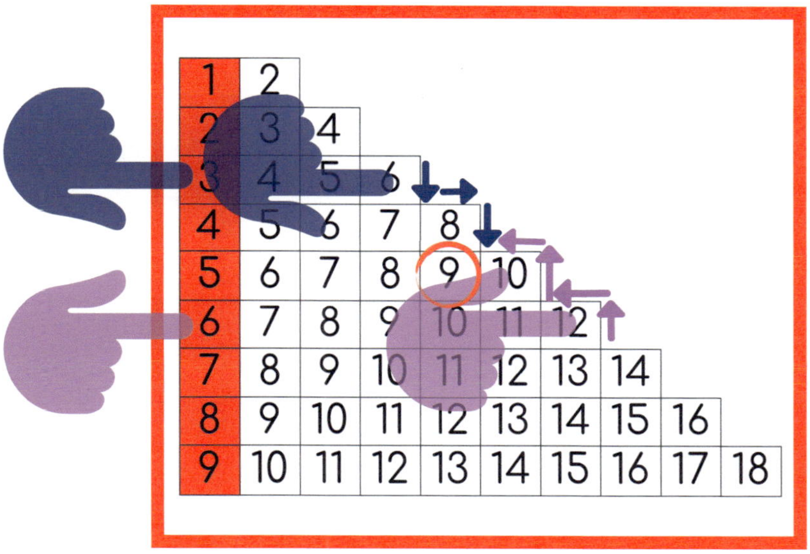

| | | |
|---|---|---|
| 1 + 7 = | 3 + 4 = | 4 + 7 = |
| 2 + 8 = | 3 + 3 = | 7 + 4 = |
| 9 + 9 = | 7 + 1 = | 8 + 6 = |
| 5 + 4 = | 6 + 1 = | 6 + 3 = |
| 8 + 4 = | 2 + 5 = | 4 + 3 = |

# Addition Working Chart 4 - Grid Paper for Sums

I know it!

I'm still practicing.

# 2.4 | Addition Working Chart 5

**Materials:**

- **Addition Working Chart 5** (Removed from the Workbook, page **A-13**; preferably laminated.)

**Directions:**

1. Prepare a basket of equations for the day by cutting apart all or some of the equations strips below.
2. Demonstrate how to use the **Addition Working Chart 5** by first choosing the equation **8 + 3 =** and recording it on the grid paper on the following page. (The example **8 + 3 =** is shown in the image below.)
3. Place one index finger on the first addend and the index finger of your other hand on the second addend.
4. Just as  we did using Addition Working Chart 4, move your index fingers as far to the right as you can, landing them on the numbers at the end of each row.
5. "Walk" your index fingers up or down the "stairs" at the same pace until they meet at either a number or a blank space. If they meet at a number, this is the sum. If they meet at a blank space, the sum is the number down one and to the left one (or the space immediately diagonally down to the left.)
6. Check the sum is correct using the **Addition Control Chart 1.**
7. Invite your student to record the sum.
8. At this point, invite your student to label the left column "I know it!" and the right column "I'm still practicing."
9. Invite your student to continue with more equations from the basket.

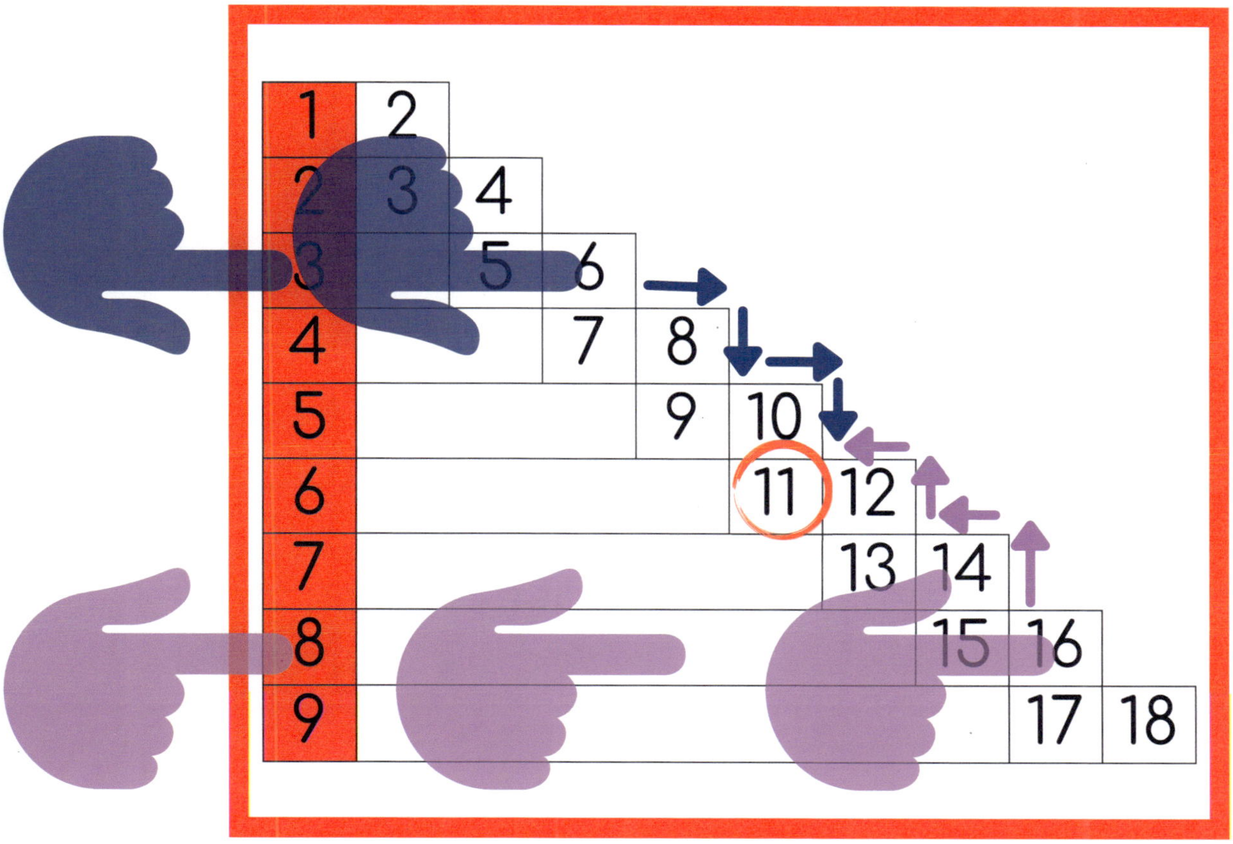


| | | |
|---|---|---|
| 5 + 9 = | 8 + 8 = | 9 + 3 = |
| 5 + 2 = | 7 + 6 = | 6 + 8 = |
| 4 + 6 = | 2 + 2 = | 8 + 3 = |

# Addition Working Chart 5 - Grid Paper for Sums

# 3.1 | Subtraction Strip Board: Introduction

**Materials:**

- **Subtraction Strip Board** (Page **A-15**, Laminated)
- **Red and Blue Number Strips** (Page **A-3,** Laminated and Cut; Note: These are the same as those used for the Addition Strip Board.)
- **Tan Unmarked Strips** (Page **A-3**, Laminated and Cut)

**Background for the Teacher:**
The **Subtraction Strip Board** is used for the *memorization* strand of Montessori math. Your student will practice "math facts" using the Strip Board. Memorization is a product of the work, not the work itself.

**Directions:**

1. Invite your student to watch as you set up the **Subtraction Strip Board** with the blue number strips on the left and the red number strips on the right like "angel wings," as you did for the **Addition Strip Board**.
2. For the tan unmarked strips, align the longest strip vertically above the number 1 at the top of the **Subtraction Strip Board,** and continue down to the shortest strip. (See image.)
3. Name the work, **"This is the Subtraction Strip Board. We will use this board to do subtraction problems."**
4. In today's lesson, you will show your student how to use the Subtraction Strip Board using a minimum of three equations. Cut apart the equations from the bottom of this page and place them in a basket or bowl.

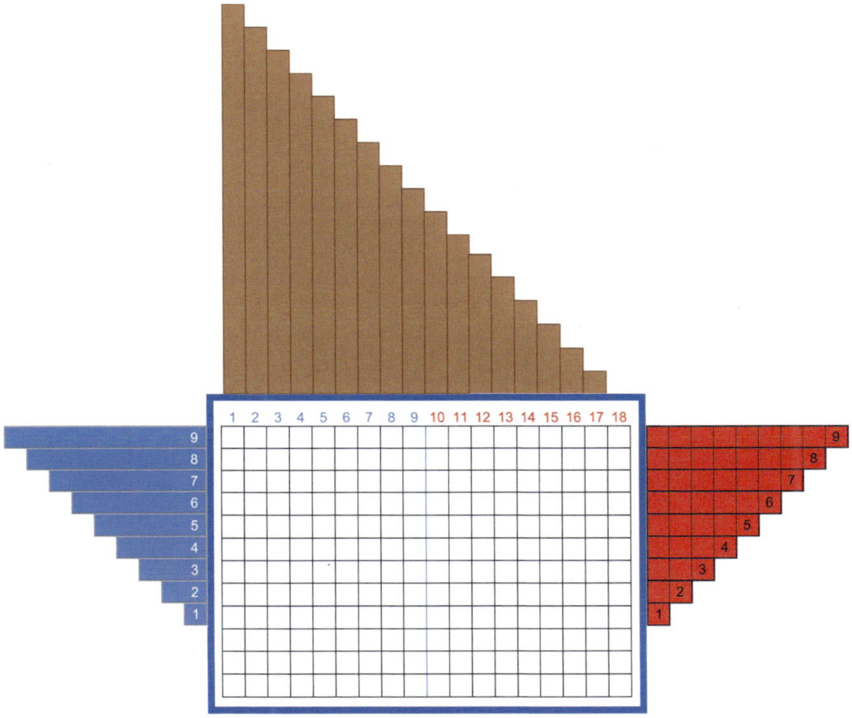


| | | |
|---|---|---|
| 9 - 6 = | 15 - 8 = | 11 - 2 = |
| 16 - 9 = | 10 - 7 = | 13 - 5 = |

5. Choose **13 - 5 =** and read it aloud as you point to each term. **"Thirteen minus five equals."** First, take the tan strip that is directly above the number 13 and use it to cover the numbers 14-18. (See image)

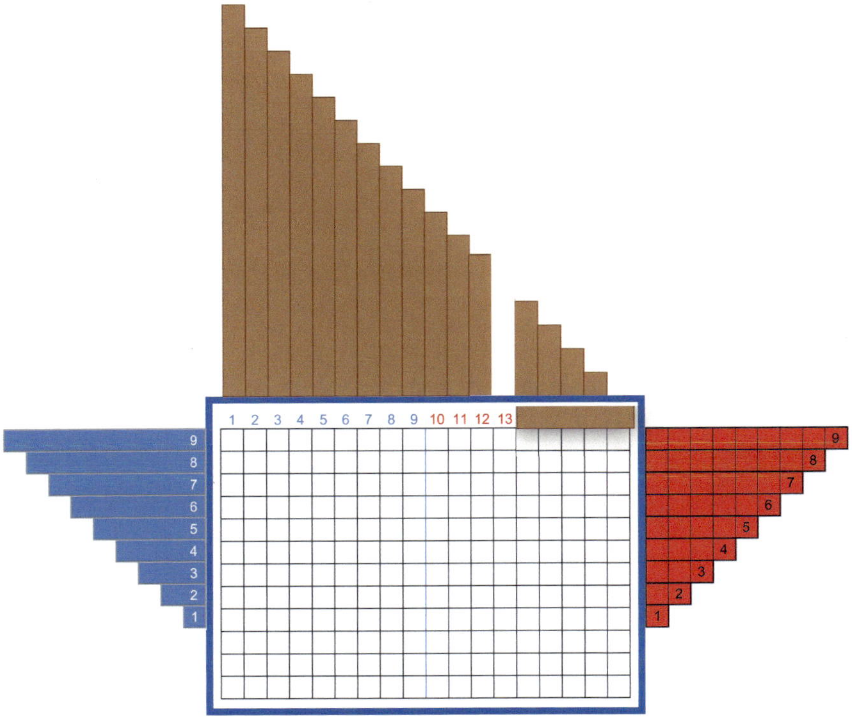

6. Then, take the blue 5 strip and align it horizontally to the tan strip, covering the numbers 9-13. Then slide the blue strip down to the top row of the squares.

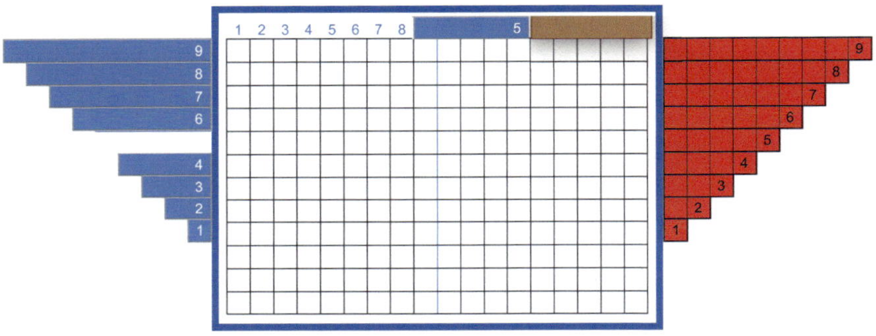

7. The **remainder** (the "answer" to the problem) is equal to the number of squares remaining to the left of the blue strip. Count these squares, then fill them with the corresponding red number strip.

8. Say, **"The answer is 8,"** as you point to the red 8 strip and the corresponding numeral 8 on the Strip Board.

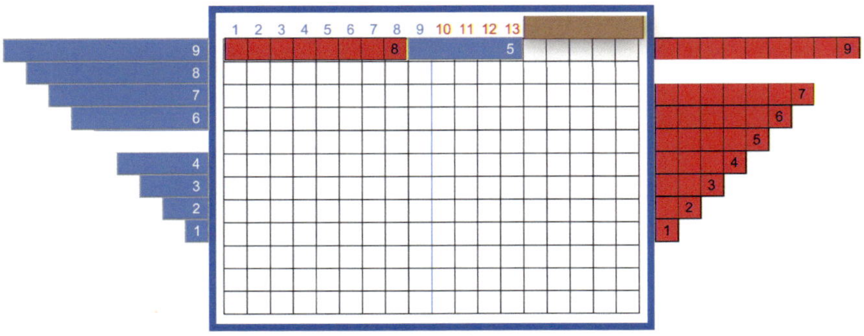

# Subtraction Tables Booklet Directions

**Materials:**

- Subtraction Strip Board (Laminated)
- Red and Blue Number Strips (Laminated and Cut)
- Tan Strips (Laminated and Cut)
- **Subtraction Control Chart 1** (Page **A-17**)

**Background for the Teacher:**
The Subtraction Tables will take several sessions to complete. Encourage the student to complete at least three tables per day to finish the booklet in 6 days. Alternatively, you could ask your student how long he would like to work (5 minutes, 20 minutes, etc.) and set a timer.

**Directions**:

1. Cut apart the pages of the Subtraction Tables Booklet that your student will be working on for this day (any from Lessons 3.2-3.6). [Alternatively, cut apart all pages from 3.2-3.6 and make the booklet ahead of time.]
2. Choose one page and read the title, **"Subtraction Table of 18."** Then point to and read aloud the equation, **"Eighteen minus nine equals."** Say, **"There is only one equation in this Subtraction Table."**
3. Show your child the process of subtraction using the Subtraction Strip Board. These are the same steps as in the previous lesson, *Subtraction Strip Board: Introduction*.
   a. Set up the Subtraction Strip Board. Use the "angel wings," with the blue strips on the left and red strips on the right. Set up the tan strips at the top of the board, aligned with the squares 1-17, with the longest strip above the 1 and the shortest strip above the 17. (See image below.)
   b. Say, **"We do not need a tan strip today, because there is not one above the number 18."**
   c. Choose the blue strip matching the minuend (9) from the left side of the board. Align it horizontally, covering the numbers on the top of the Strip Board, so that the right side of the strip covers the 18. Then move the blue strip down to the first row.
   d. Count the number of empty squares to the left of the blue strip and choose the corresponding red number strip (9). Point to the red number strip and the red number 9 at the top and say, "This is the remainder. This is the answer. Nine. 18 - 9 = **9**."
   e. Invite your student to check the answer using the **Subtraction Control Chart 1** then record the answer in the booklet.
   f. Invite your student to complete the Subtraction Table of 17- in the same manner, setting up the board with the tan strip covering the number 18 now.
   g. Complete a minimum of 3 tables per day to finish this exercise in 6 days. (This is just a suggested pace!)
4. Invite your student to staple all of the booklet pages together in order, with the cover as the first page.

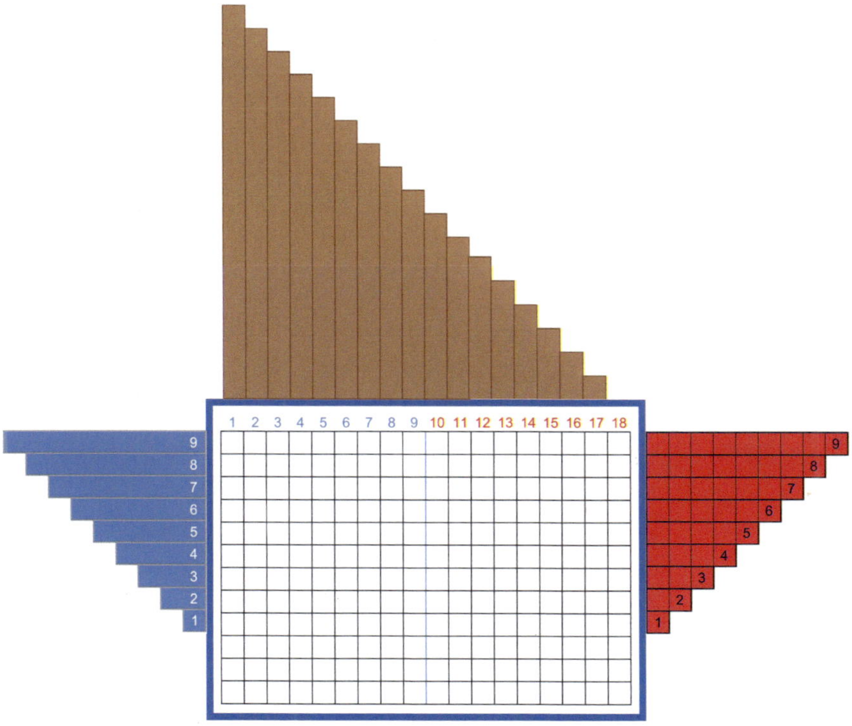

subtraction table
of
18 -

18 - 9 = ______

subtraction table
of
17 -

17 - 9 = ______

17 - 8 = ______

subtraction table
of
16 -

16 - 9 = ______

16 - 8 = ______

16 - 7 = ______

subtraction table
of
15 -

15 - 9 = ______

15 - 8 = ______

15 - 7 = ______

15 - 6 = ______

subtraction table
of
14 -

14 - 9 = ________
14 - 8 = ________
14 - 7 = ________
14 - 6 = ________
14 - 5 = ________

subtraction table
of
13 -

13 - 9 = ________
13 - 8 = ________
13 - 7 = ________
13 - 6 = ________
13 - 5 = ________
13 - 4 = ________

subtraction table
of
12 -

12 - 9 = ________
12 - 8 = ________
12 - 7 = ________
12 - 6 = ________
12 - 5 = ________
12 - 4 = ________
12 - 3 = ________

subtraction table
of
11 -

11 - 9 = ________
11 - 8 = ________
11 - 7 = ________
11 - 6 = ________
11 - 5 = ________
11 - 4 = ________
11 - 3 = ________
11 - 2 = ________

subtraction table
of
10 -

10 - 9 = ______
10 - 8 = ______
10 - 7 = ______
10 - 6 = ______
10 - 5 = ______
10 - 4 = ______
10 - 3 = ______
10 - 2 = ______
10 - 1 = ______

subtraction table
of
9 -

9 - 9 = ______
9 - 8 = ______
9 - 7 = ______
9 - 6 = ______
9 - 5 = ______
9 - 4 = ______
9 - 3 = ______
9 - 2 = ______
9 - 1 = ______

subtraction table
of
8 -

8 - 8 = ______
8 - 7 = ______
8 - 6 = ______
8 - 5 = ______
8 - 4 = ______
8 - 3 = ______
8 - 2 = ______
8 - 1 = ______

subtraction table
of
7 -

7 - 7 = ______
7 - 6 = ______
7 - 5 = ______
7 - 4 = ______
7 - 3 = ______
7 - 2 = ______
7 - 1 = ______

subtraction table
of
6 -

6 - 6 = ______

6 - 5 = ______

6 - 4 = ______

6 - 3 = ______

6 - 2 = ______

6 - 1 = ______

subtraction table
of
5 -

5 - 5 = ______

5 - 4 = ______

5 - 3 = ______

5 - 2 = ______

5 - 1 = ______

subtraction table
of
4 -

4 - 4 = ______

4 - 3 = ______

4 - 2 = ______

4 - 1 = ______

subtraction table
of
3 -

3 - 3 = ______

3 - 2 = ______

3 - 1 = ______

subtraction table
of
2 -

2 - 2 = ______

2 - 1 = ______

subtraction table
of
1 -

1 - 1 = ______

______'s

Subtraction
Tables
Booklet

# 3.7 | Subtraction Strip Board: Basket of Equations

**Materials**

- Subtraction Strip Board (Laminated)
- Red and Blue Number Strips (Laminated and Cut)
- Tan Strips (Laminated and Cut)
- Subtraction Control Chart 1 (Page **A-17**)

**Directions:**

1. Prepare the basket of equations (or invite the student to prepare the basket of equations) by cutting apart the equations on this page and placing them in a small basket or bowl.
2. Set up the Subtraction Strip Board as before. (See Introduction lesson, if needed.)
3. The student should take an equation out of the basket, record it on the grid paper, then perform the subtraction using the strip board. (See Introduction lesson if you need a refresher.)
4. The student should then record the difference on the grid paper.
5. Continue for as many equations as your student chooses. (Hint: Set a goal of a certain number of equations or working for a certain number of minutes if your student needs further motivation.)
6. When the child finishes the work, he or she should check the equations using the Subtraction Control Chart 1. If there are any errors, the student can choose to redo those problems with the Strip Board.

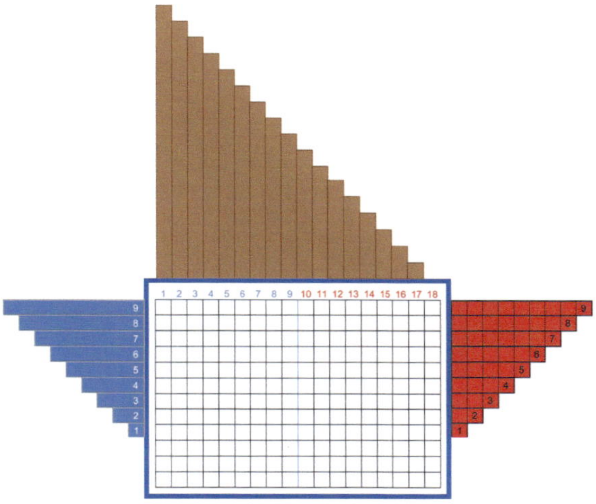

| | | |
|---|---|---|
| 8 - 3 = | 6 - 4 = | 5 - 2 = |
| 4 - 1 = | 11 - 9 = | 9 - 3 = |
| 8 - 2 = | 12 - 6 = | 12 - 4 = |
| 13 - 4 = | 13 - 8 = | 14 - 9 = |
| 14 - 7 = | 15 - 6 = | 16 - 8 = |
| 17 - 8= | 3 - 2 = | 5 - 3 = |

# Subtraction Strip Board - Basket of Equations

# 3.8 | Subtraction Strip Board: Unit Analysis of 9

**Materials:**

- Red and Blue Colored Pencils
- Subtraction Control Chart 1

**Directions:**

1. Say, **"Today we are going to analyze the number 9."** The board is already prepared with 9 as the minuend. (Point to the tan strip covering the numbers to the right of 9 at the top of the board.)
2. Using colored pencils, color in the subtrahend using blue, starting at the grid space directly below the minuend number and counting the proper number of spaces to the left that correspond to the subtrahend.
3. Fill in the remaining grid spaces to the left of the blue subtrahend with red colored pencil to show the remainder. (See sample.)
4. Record the remainder on the space provided.
5. Continue performing the subtraction equations in this manner, working in each subsequent row and recording equations.
6. When the student completes all of the unit analysis equations for one number, he should check his work with the **Subtraction Control Chart 1.**
7. Continue in this manner for the remaining **Unit Analysis** problems, working at a pace that is comfortable for your student.

Sample:

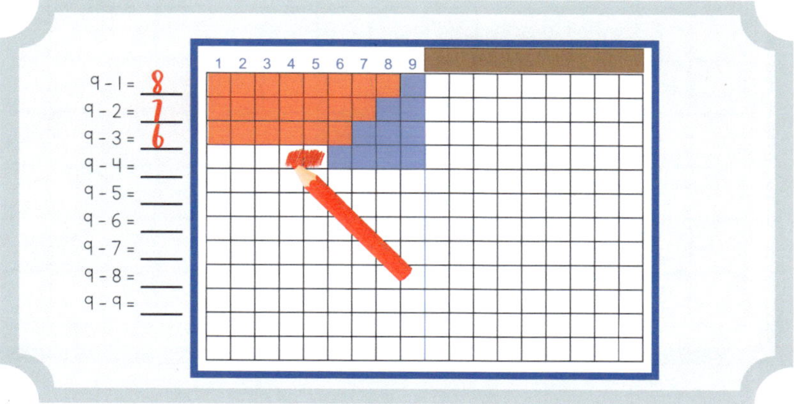


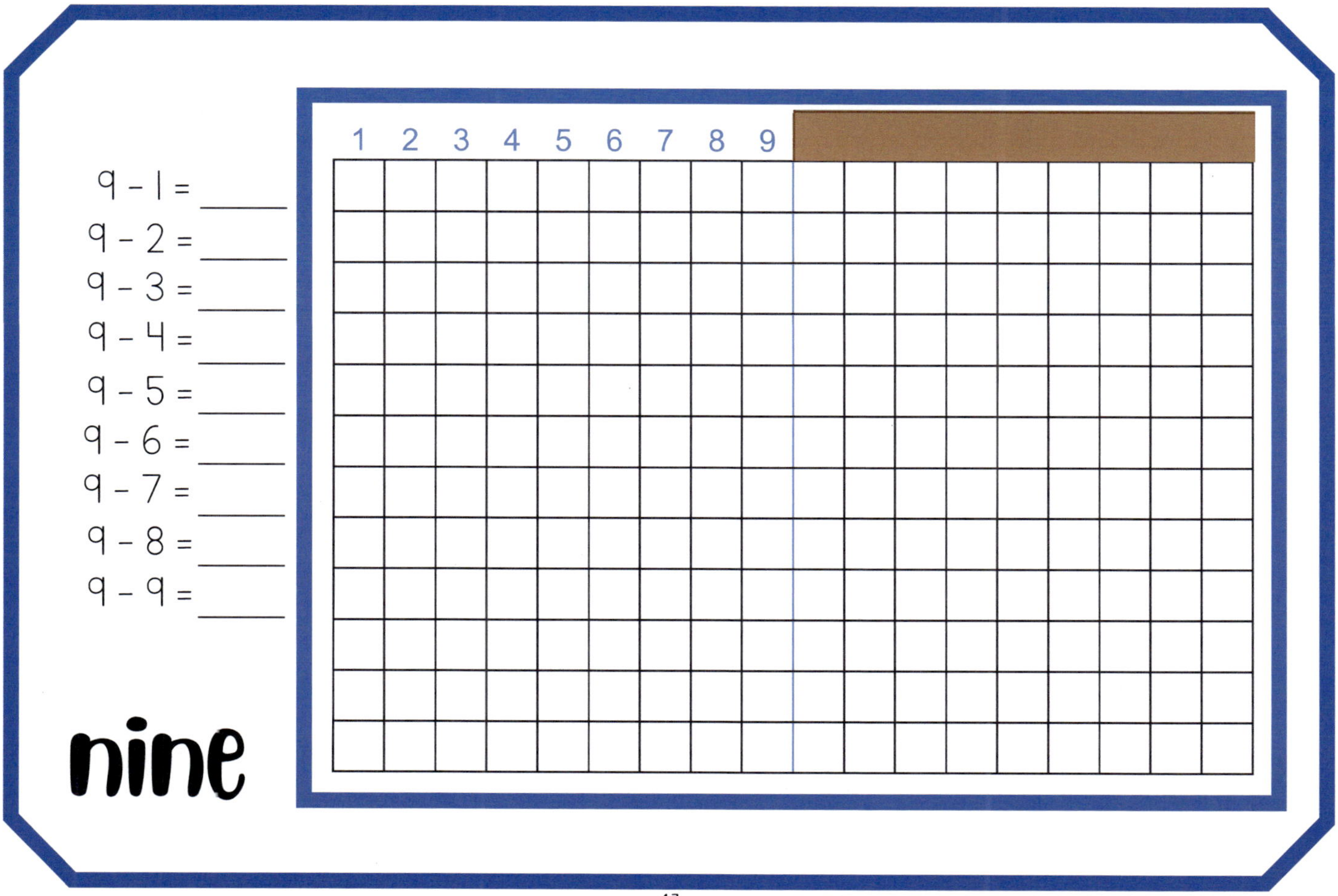

# 3.9 | Subtraction Strip Board: Unit Analysis of 8 & 7

1 2 3 4 5 6 7 8

8 - 1 = ____
8 - 2 = ____
8 - 3 = ____
8 - 4 = ____
8 - 5 = ____
8 - 6 = ____
8 - 7 = ____
8 - 8 = ____

**eight**

1 2 3 4 5 6 7

7 - 1 = ____
7 - 2 = ____
7 - 3 = ____
7 - 4 = ____
7 - 5 = ____
7 - 6 = ____
7 - 7 = ____

**seven**

# 3.10 | Subtraction Strip Board: Unit Analysis of 6 & 5

6 - 1 = ______
6 - 2 = ______
6 - 3 = ______
6 - 4 = ______
6 - 5 = ______
6 - 6 = ______

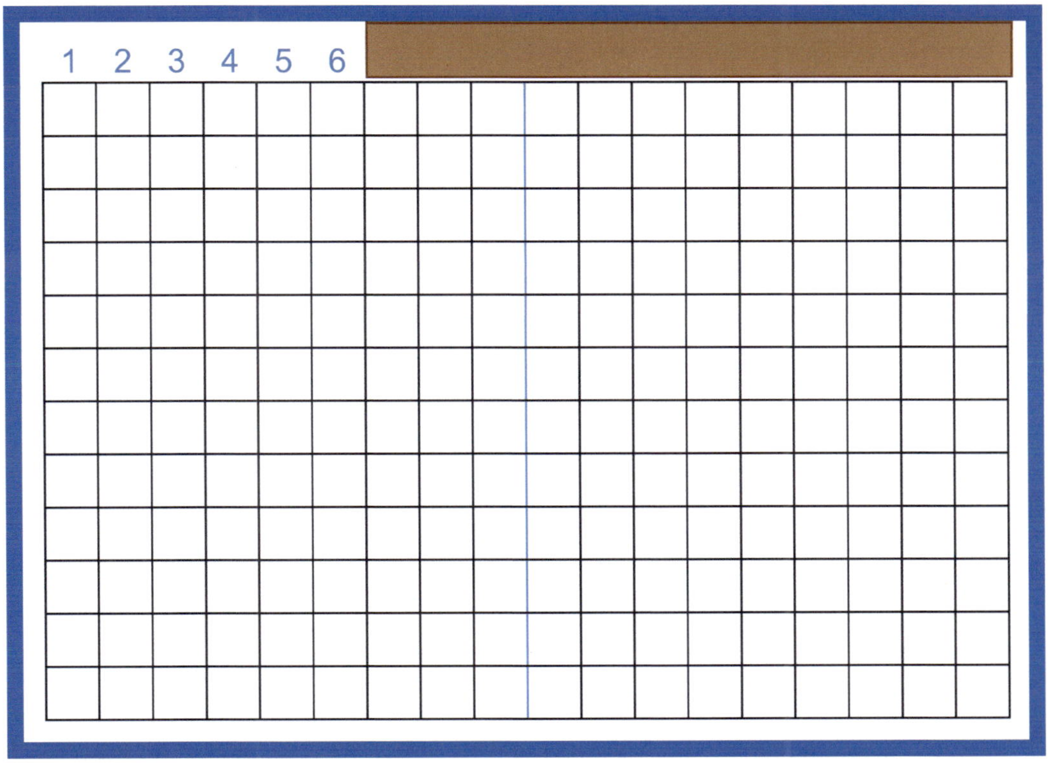


**six**

5 - 1 = ______
5 - 2 = ______
5 - 3 = ______
5 - 4 = ______
5 - 5 = ______

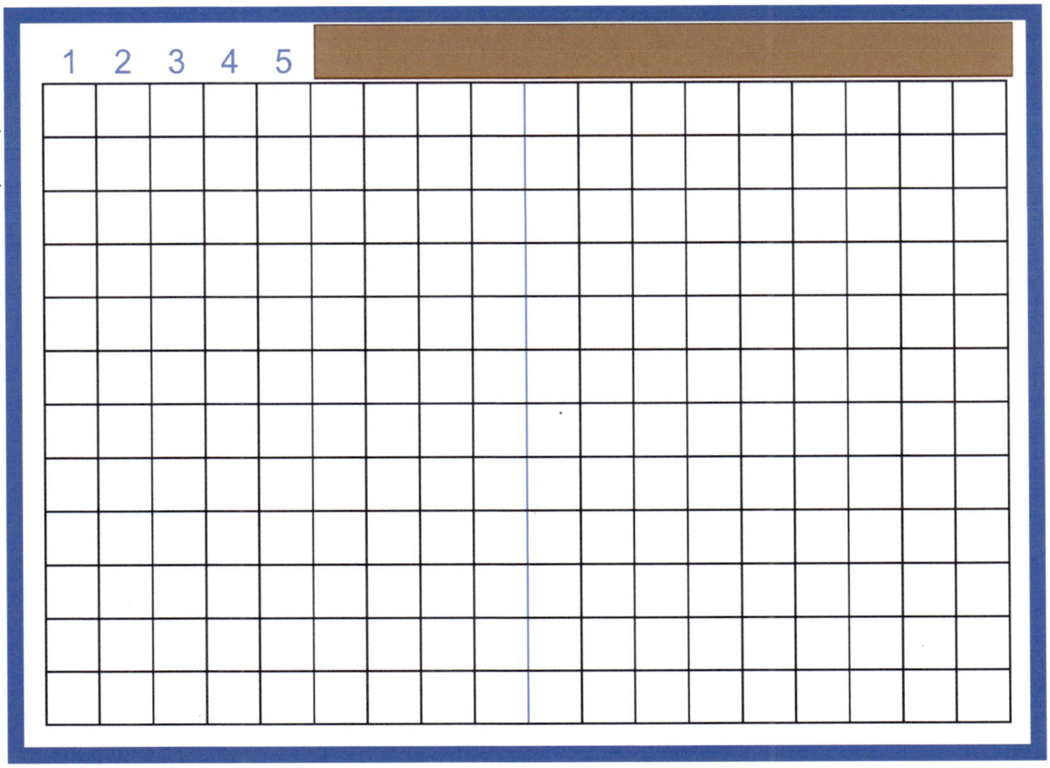


**five**

4 - 1 = ______

4 - 2 = ______

4 - 3 = ______

4 - 4 = ______

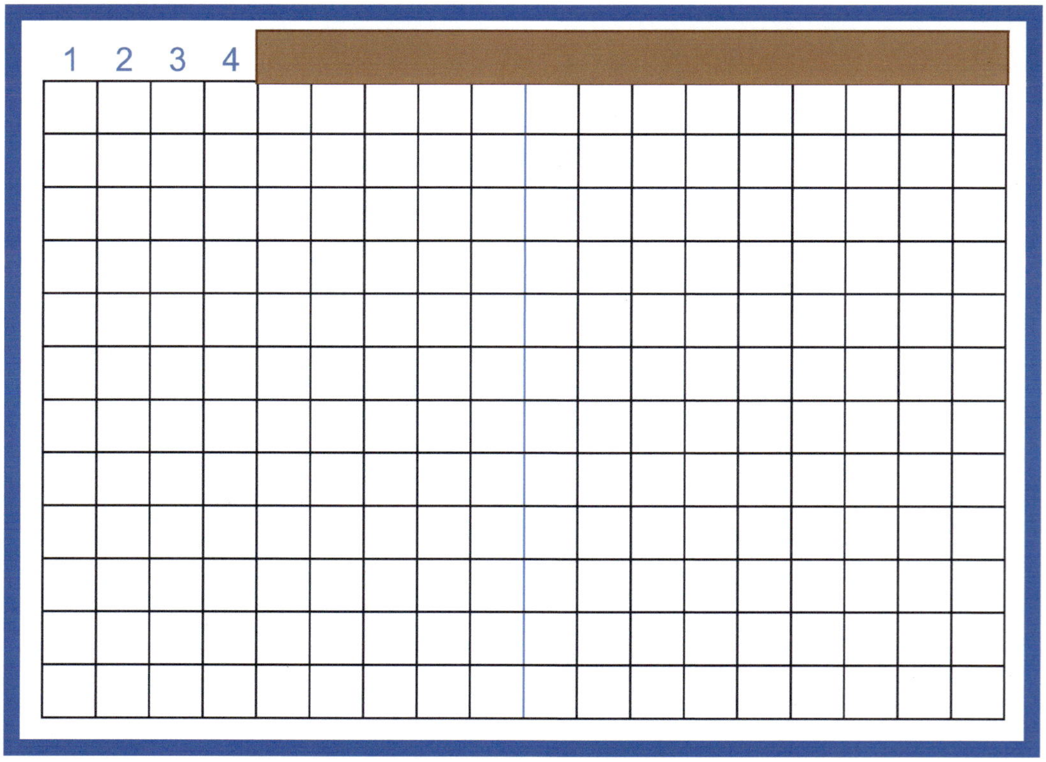


**four**

3 - 1 = ______

3 - 2 = ______

3 - 3 = ______

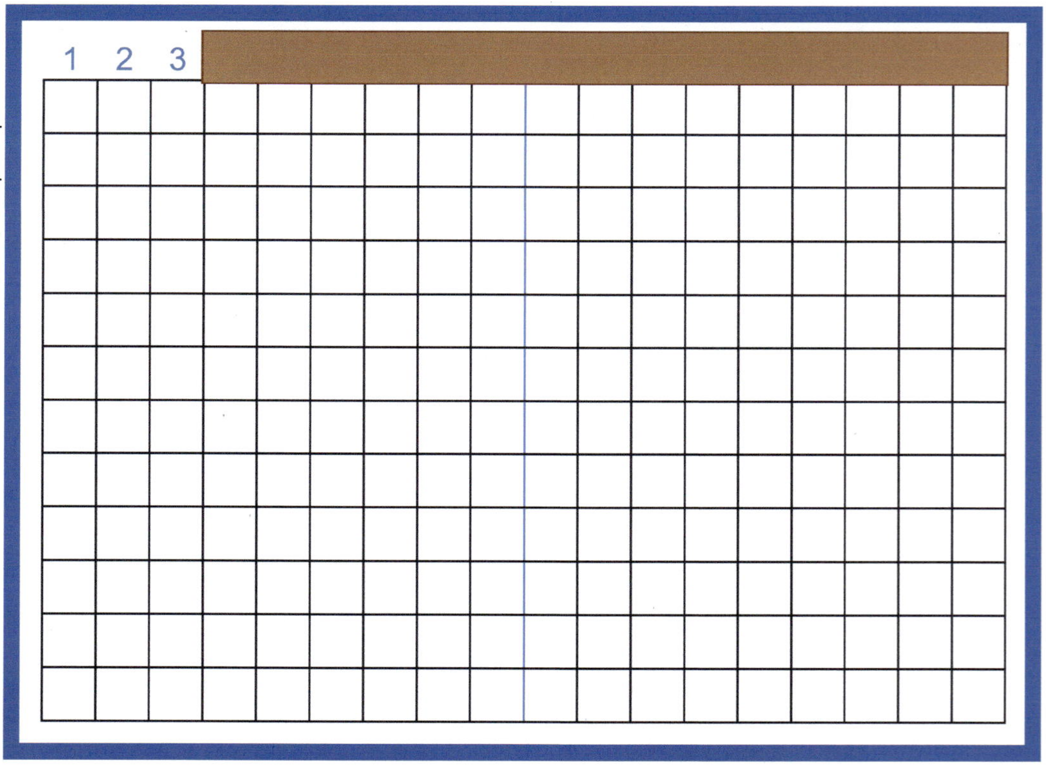


**three**

2 - 1 = ______

2 - 2 = ______

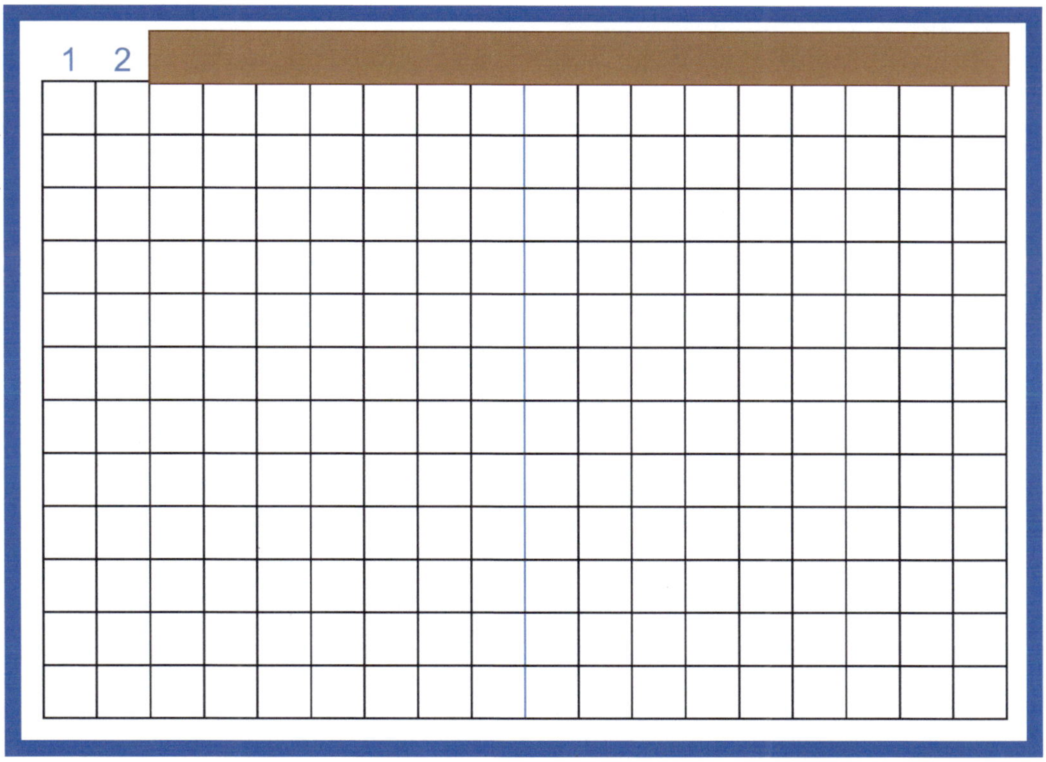


two

1 - 1 = ______

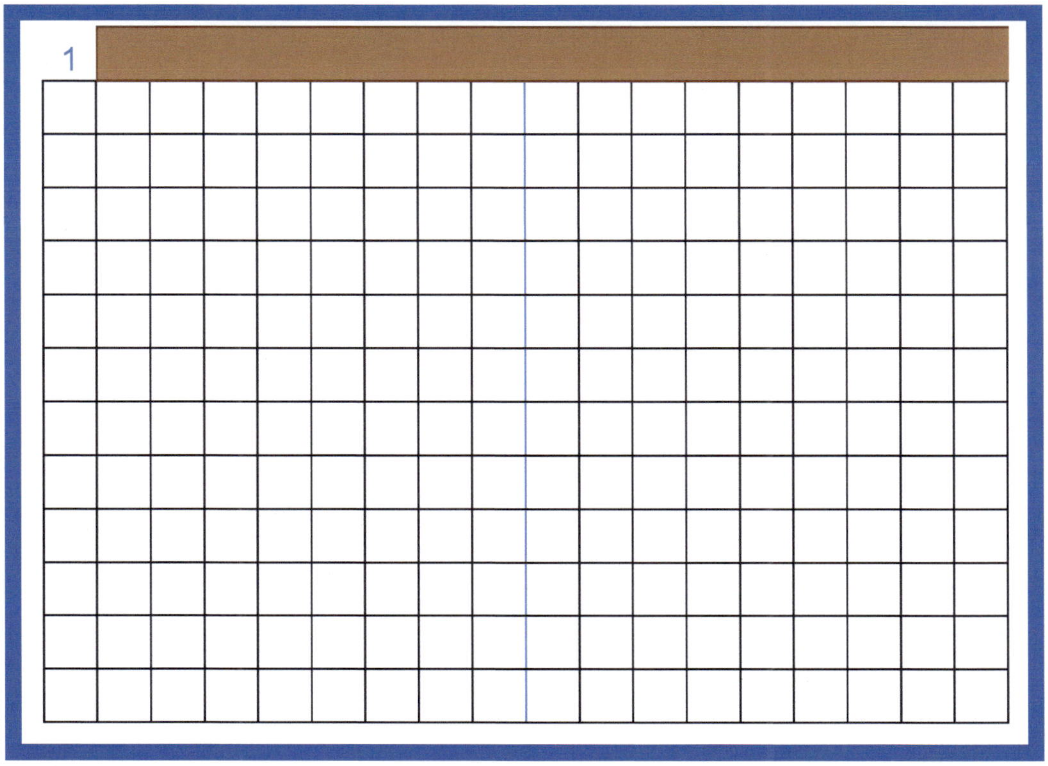


one

# Bead Houses Introduction

**Background for Teacher:**
The Bead Houses use elements of addition and subtraction as the child fills in all possible combinations of numbers that add to each sum from one to ten.
The goal is to analyze each sum for its component parts. The student will see that equations are repeated by reversing the addends with a beautiful visual of colored beads. The student will also keep a special notebook for these bead houses (see page **63**).

**Preparation:**
1. Prepare the special Bead House Notebook (see page **63.**)
2. Cut out the bead strip corresponding to the first bead house you will show your child how to complete: the Bead House of 4. (The beads for the House of 4 are found in Lesson 4.3. The Bead House of 4 is found in lesson 4.2.)

**Directions:**
1. Invite the student to write the House Number in the circle on the "roof" of the yellow Bead House of Four - found in Lesson 4.2. (The Bead Houses and Beads are colored as follows: Red = 1, Green = 2, Pink = 3, Yellow = 4, Light Blue = 5, Purple = 6, White = 7, Brown = 8, Dark Blue = 9, Gold = 10.)
2. Next, show your child how to do the work using the **House of 4**. (This is a good one to demonstrate, because it doesn't have too few or too many equations.)
3. Cut apart the beads corresponding to the house and lay them face up to the right of the workbook. (The beads for the House of 4 are in Lesson 4.3.)
4. Choose the **bead bar that matches the house** (the bead bar of 4) and place it in the **top left space**. Point to the bead bar of four and count each bead aloud as you point to each one.
5. Next, choose the bead bar of three (*one less than the previous bead bar*) and place it in the first column of the second row. **Count the beads**. ***"One, two, three."*** Then say, **"I need this row to have a total of four beads. I will place the 3-bar under the bead bar of four to see how many more beads I need." Align the bead bar strips** and show that you need one more bead. In the second row of the bead house, place the bead bar of three in the left column and the bead bar of one in the right column. **Check your work by counting** starting with the pink bead bar, **"One two, three,"** and continue counting the red bead bar, **"four."**
6. Continue in the next row with the bead bar of two. Show your child how you place the bead bar of two under the bead bar of four to determine the missing bead bar to add to the right column.
7. The last row will be zero and four.
8. **Record the equations** on the prepared special **bead bar booklet** paper.
9. Allow your child to work through the Bead Houses at a comfortable pace, setting quantity *("I'll do three Bead Houses today!")* or time goals *("I will work for ten minutes!")* as needed. Or you can do one "Lesson" a day.

**Extension/Option:**
Some guides may choose to remove the bead houses, beads, and bead bar booklet from the workbook prior to introducing this work to their student(s) to focus on math over cutting/pasting or to prepare shelf work.

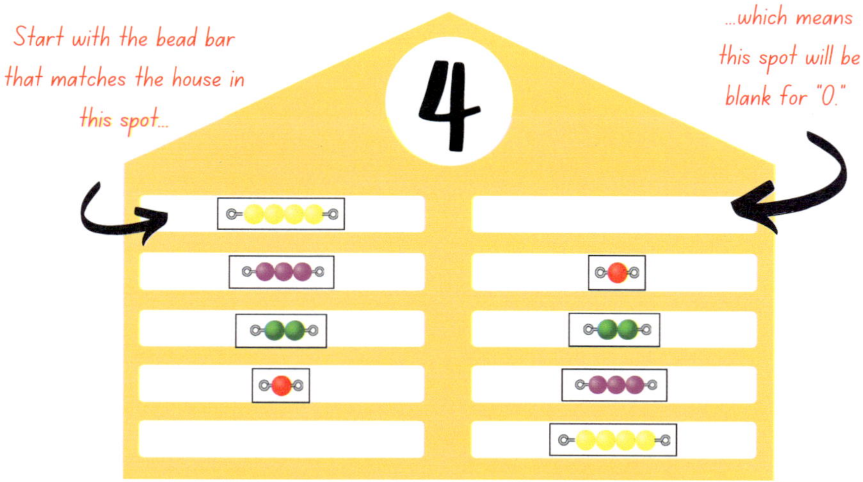


*Bead House - completed*

*Bead House Booklet page - completed*

# 4.1 | Bead Houses of 1 & 2

Beads for House of 1

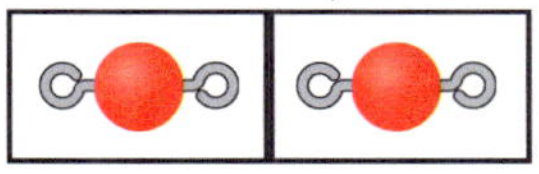

Beads for House of 2

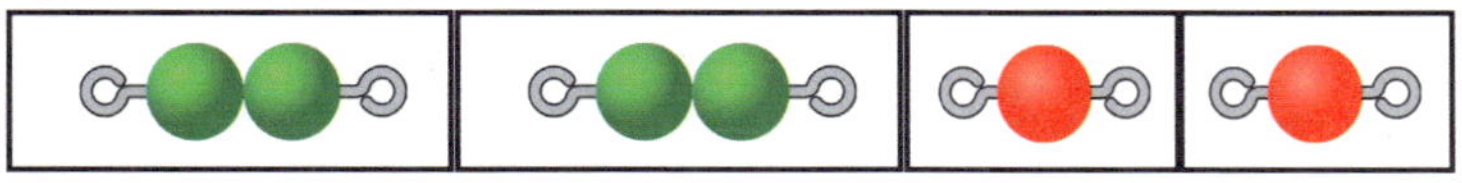

# 4.2 | Bead Houses of 3 & 4

The Beads for the House of 3 and the House of 4 are on the next page, with the House of 5.

# 4.3 | Bead House of 5

Beads for House of 5

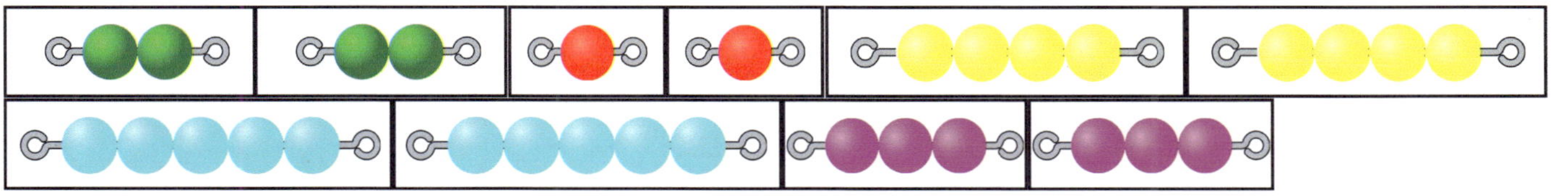

Beads for House of 4

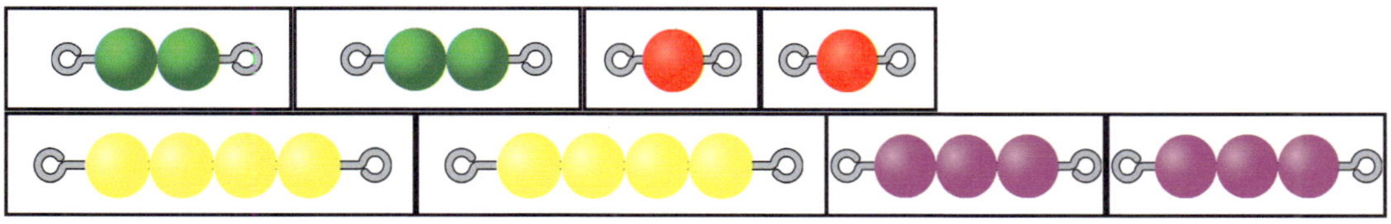

Beads for House of 3

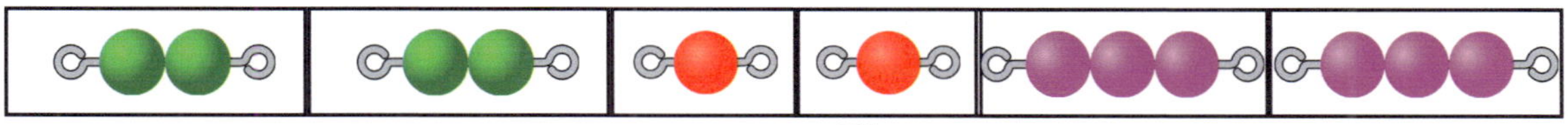

# 4.4 | Bead House of 6

Beads for House of 6

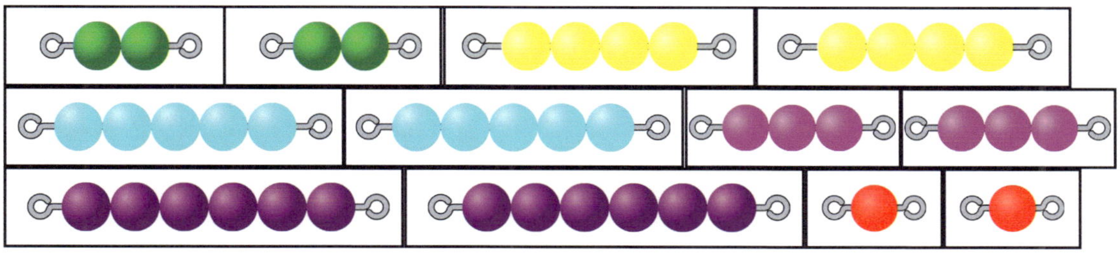

# 4.5 | Bead House of 7

Beads for House of 7

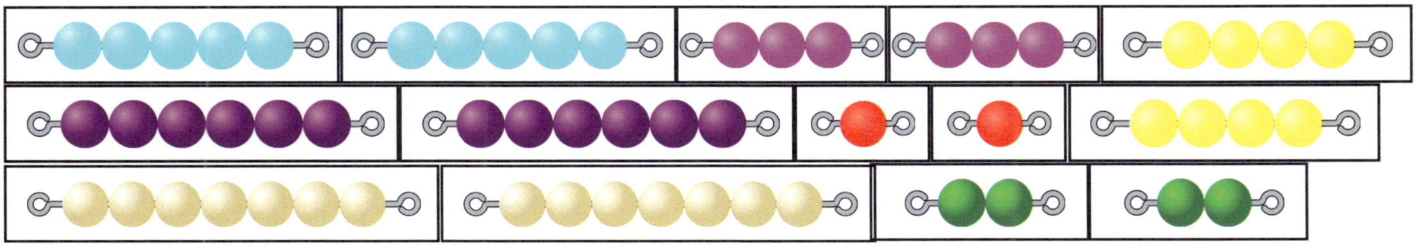

# 4.6 | Bead House of 8

Beads for House of 8

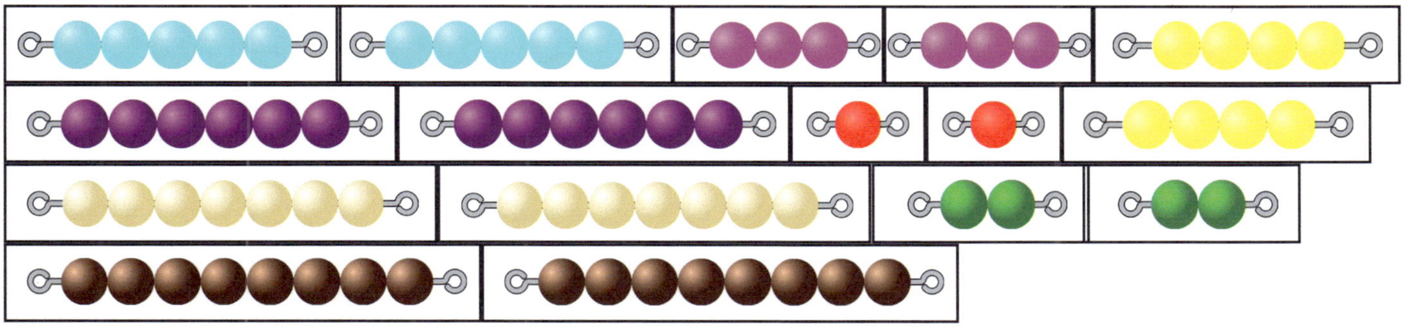

# 4.7 | Bead House of 9

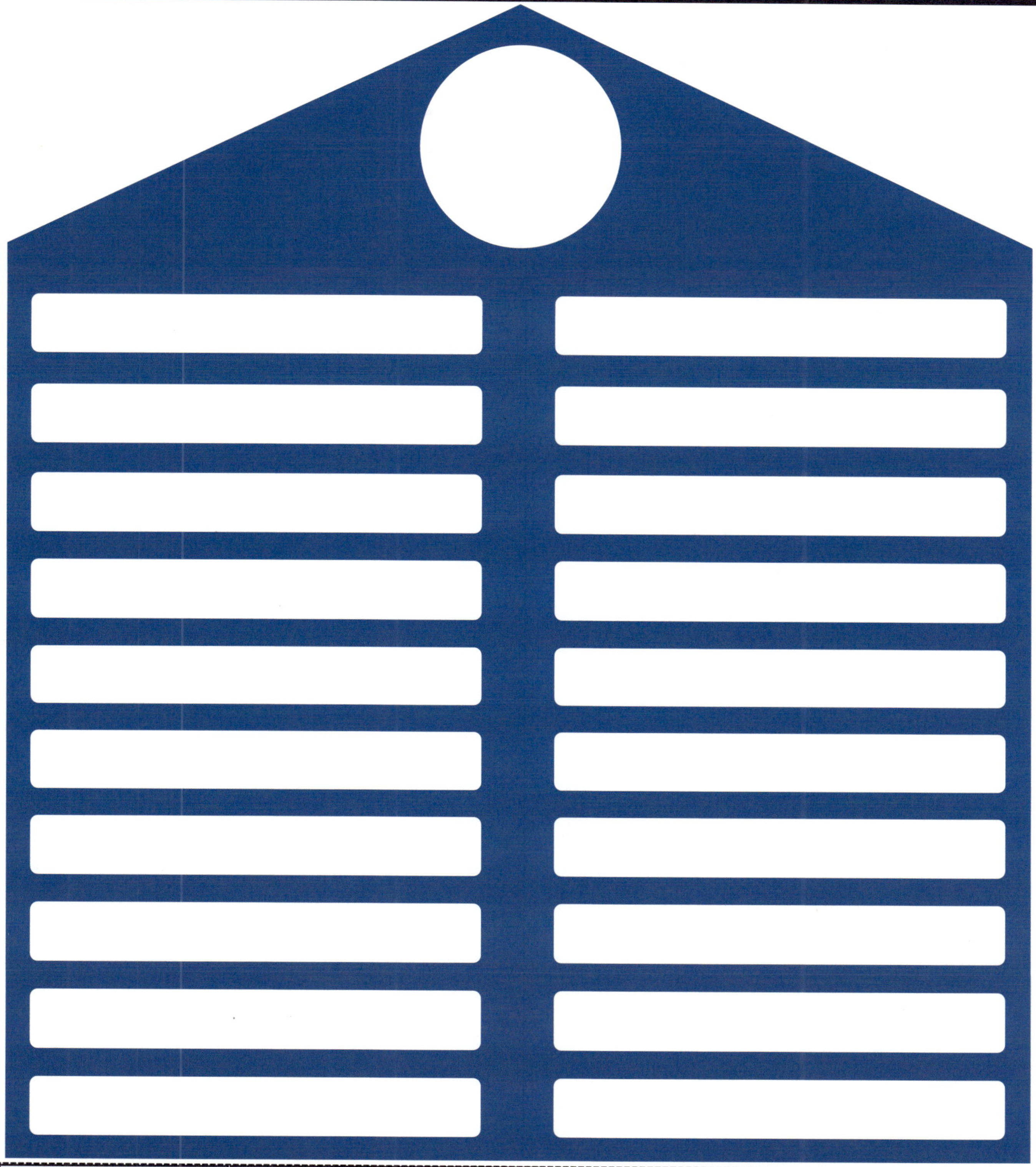

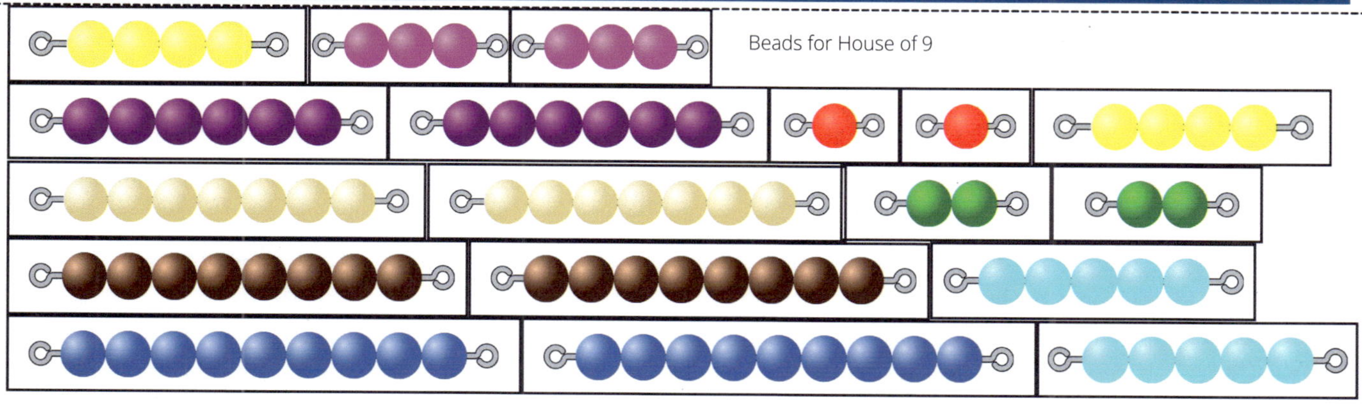

# 4.8 | Bead House of 10

Beads for House of 10

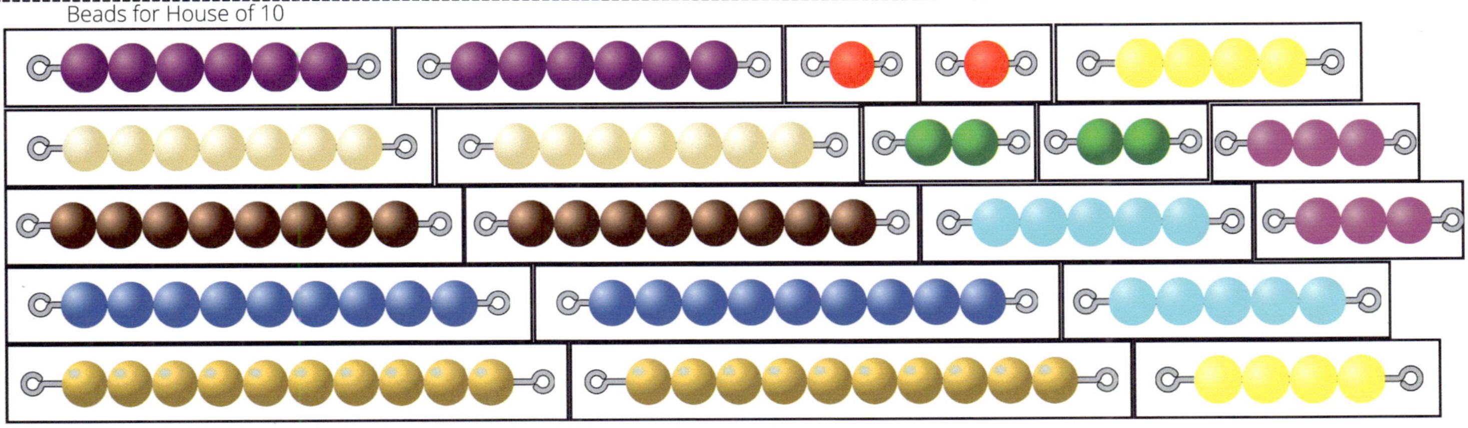

# Bead House Notebook Pages

**Directions**:

1. Cut out each notebook page to record the "houses" of equations as your student completes each bead house in Lessons 4.1-4.8.
2. Staple together in order when all pages are complete, with the **Cover Page** on top, followed by the **House of 1, 2**, etc., and the **House of 10** on the bottom.

**Cover Page**

Put this page on top, followed by the bead house notebook pages 1-10. Then, staple on left "roof" as indicated.

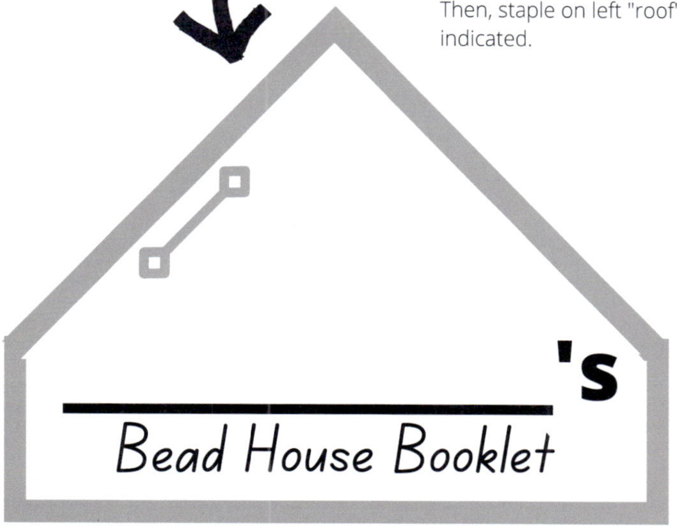

4
house of 4

6
house of 6

5
house of 5

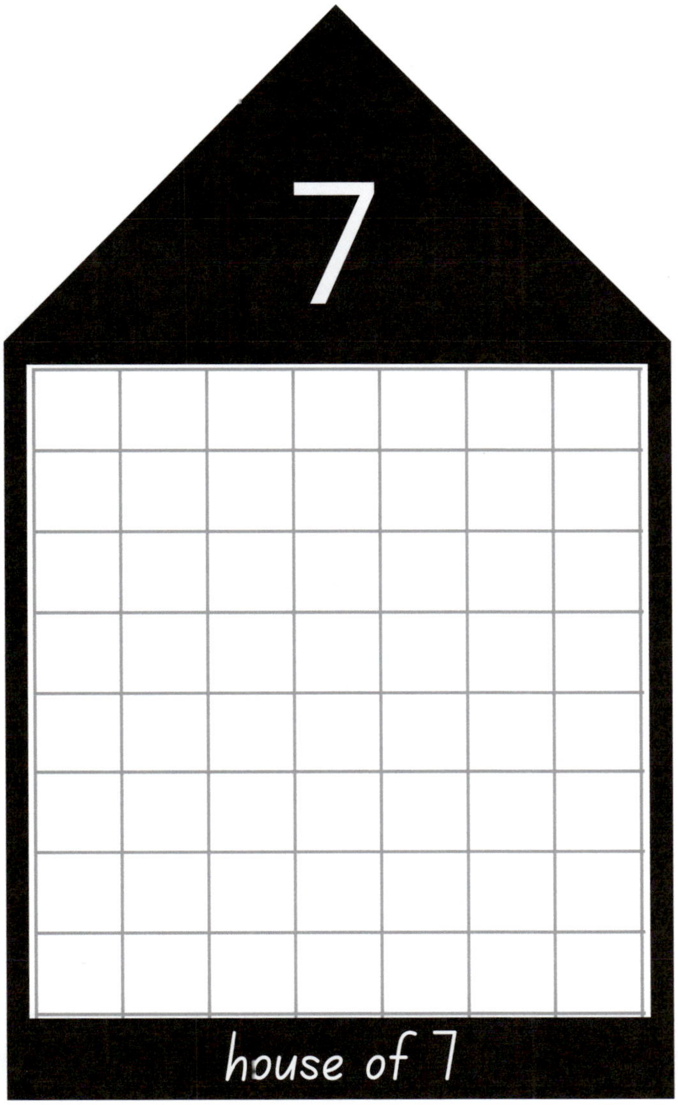
7
house of 7

8
house of 8

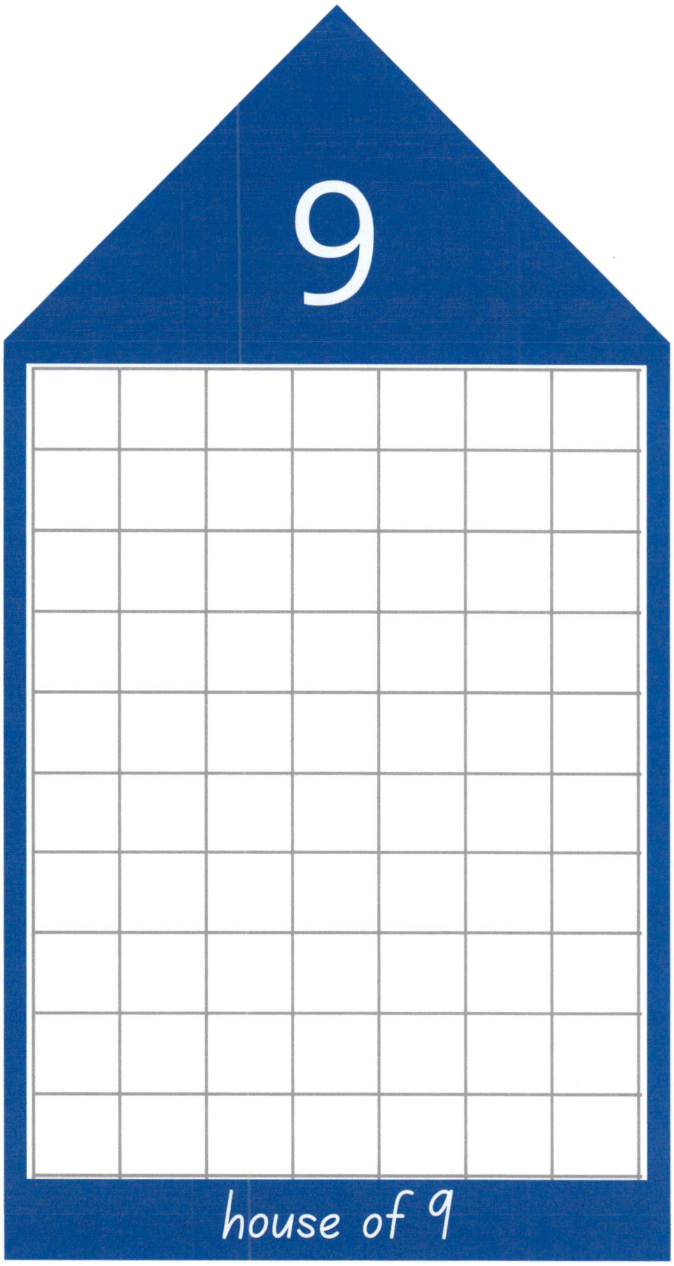
9
house of 9

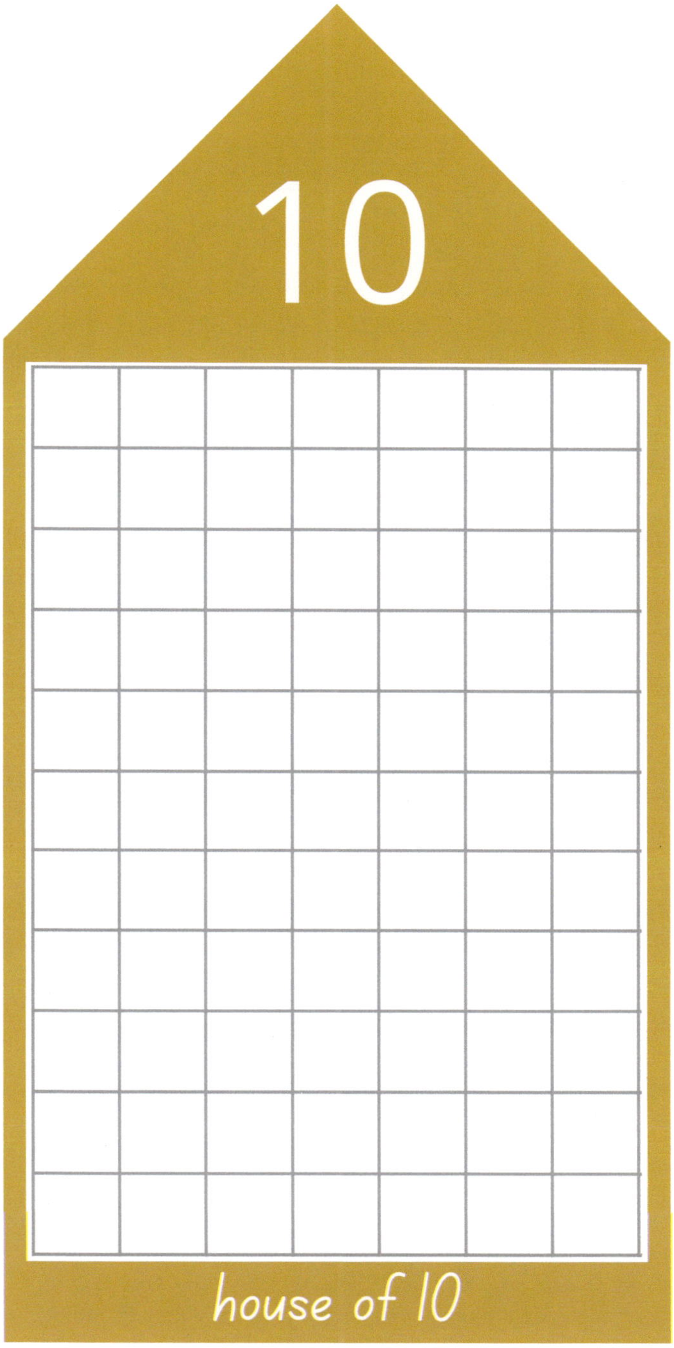
10
house of 10

# 5.1 | Stamp Game Introduction: Part 1

**Materials:**

- Stamp Game Printables cut and placed into an appropriate container OR a **wooden Stamp Game** (See Page **A-19**)
- *Optional* **Golden Bead Material** or 3D options from the *Montessori Math Workbook: Primary Book 2*: use One 3-D Thousand Cube, One Hundred Square, One Ten Bar, and One Unit Bead

**Background for Teacher:**
The Stamp Game is the next step in the arithmetic operations and decimal system sequence of math work after the operations with the Golden Bead material, which was covered in the *Montessori Math Workbook: Primary Book 2.* The stamps take the place of the Golden Beads. The child is moving from concrete math work to more abstract math work. The stamp for 1000, for instance, is the same size as the stamp for 1, yet they represent vastly different quantities.

A major goal of the Stamp Game is to give the child many opportunities for exchanges. We will practice one problem without exchanges and one problem with exchanges each time we teach a new operation, but most (if not all) problems the child does on her own will involve exchanges.

The stamps also serve an important role in preparing the child for algebra and geometry, particularly in the multiplication and division problems. When laying out the answer for a multiplication problem, the student may start to notice the natural geometric pattern that is a result of the multiplication. At this stage, there is no need to point out any connections *for* the child, but know that your student is receiving more math background with this stamp game that further prepares the child's mind for future arithmetic endeavors.

We will start by taking a couple of lessons to get to know the stamp game. In this first lesson, your student will learn about the *stamps*, the *skittles*, and the *discs*.

***OPTIONAL* Part 1**
**(Optional. Only do Part 1 if you have the physical Golden Bead Material or DIY 3-D Golden Bead Material from *Montessori Math Workbook 2. If you do not have the physical materials, move directly to Part 2 on the next page.*):**

**Materials:**

- Physical Golden Bead Material: one thousand cube, one hundred square, one ten bar, one unit bead
- Stamp Game: one of each value of stamp (1000, 100, 10, 1)

**Directions:**

1. Set out thousand cube, one hundred square, one ten bar, and one unit bead on the table. Point to each one and say, **"This is one thousand. This is one hundred. This is one ten. This is one unit."**
2. Next, show your child the Stamp Game you have prepared. (Note: See page **A-19** for storage suggestions.) Say, **"This is the Stamp Game."**
3. Take out the 1000 stamp and say, **"This stamp represents one thousand."** Then place the stamp underneath the one thousand cube. Do the same for 100, 10, and 1.
4. Say, **"Close your eyes. I am going to mix up these stamps and see if you can put them back where they go."** Mix up the stamps and put them in front of your student. **"Open your eyes and put them back under the correct amount of beads."**
5. If your child made any mistakes, stop here, put the work away and repeat this lesson another day.
6. Repeat the exercise, but mix up just the Golden Beads.
7. Repeat the exercise, but mix up the Golden Beads and the Stamps. (Note: Make sure the student puts the beads in order as they were before: 1000, 100, 10, 1.)

# 5.2 | Stamp Game Introduction: Part 2

**Part 2 Directions (You will do this part even if you completed Part 1):**

1. Cut out the pictures of the Golden Bead Materials.
2. Place the 1000 cube on the left, followed by the 100 square, the 10 bar, and the unit bead.
3. Review the quantity of each of Golden Beads (1000, 100, 10, 1.) Say, **"This cube has 1000 beads." "This square has 100 beads." "This bar has ten beads." "This is one bead."**
4. Show your child the Stamp Game you have purchased or prepared. (Note: See page **A-19** for storage suggestions.) Say, **"This is the Stamp Game."**
5. Take out the 1000 stamp and say, **"This stamp represents one thousand."** Then place the stamp underneath the one thousand cube. Do the same for 100, 10, and 1.
6. Say, **"Close your eyes. I am going to mix up these stamps and see if you can put them back where they go."** Mix up the stamps and put them in front of your student. **"Open your eyes and put them back under the correct amount of beads."**
7. If your child made any mistakes, stop here, put the work away and repeat this lesson another day.
8. Repeat the exercise, but mix up just the Golden Beads.
9. Repeat the exercise, but mix up the Golden Beads and the Stamps. (Note: Make sure the student puts the beads in order as they were before (1000, 100, 10, 1.)
10. Glue down in place on the mat.

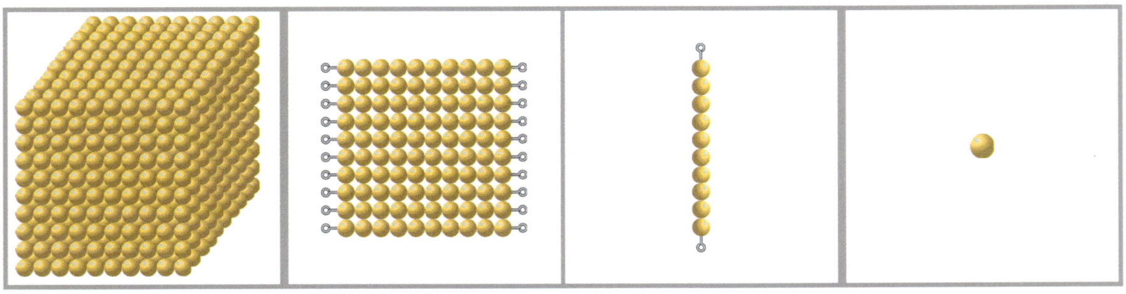

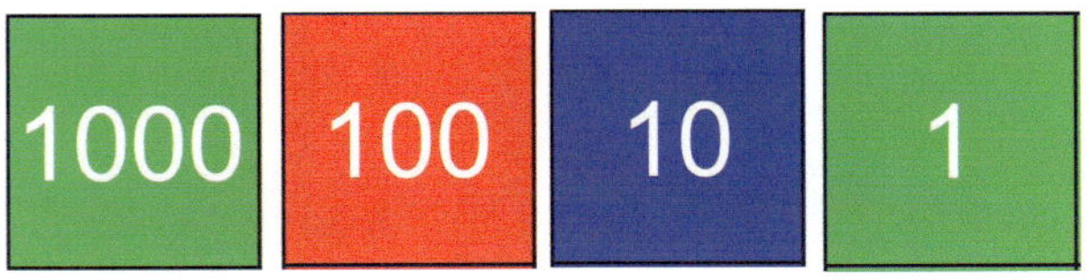

# 5.3 | Stamp Game: Formation of Numerals

**Directions:**
Invite your student to match the numeral with the quantity of stamps and then draw a line from the numeral to the quantity.

475

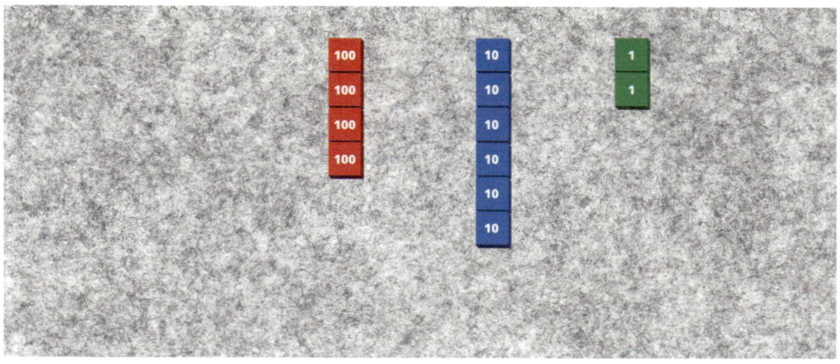


1382

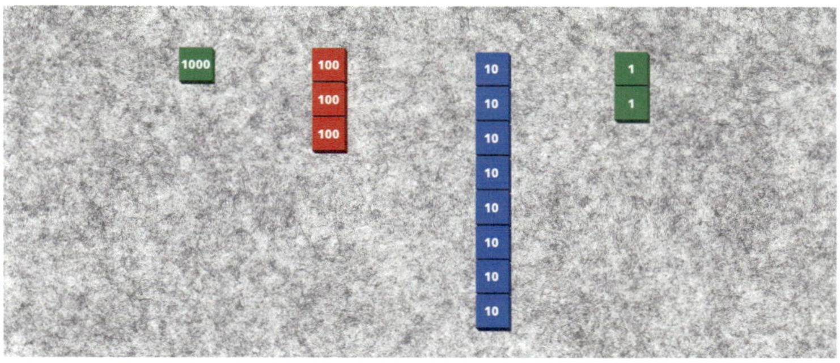


1238

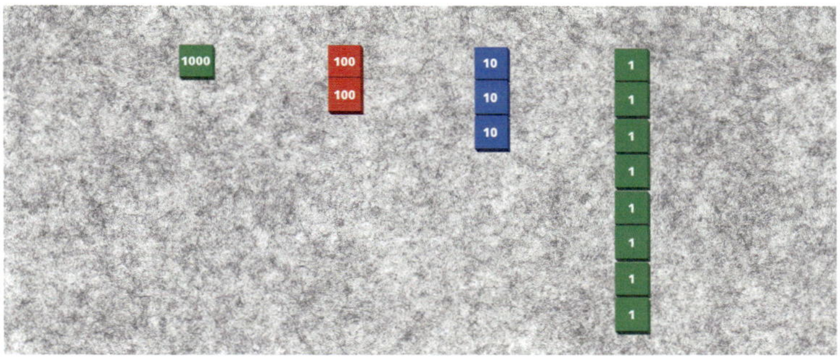


1823

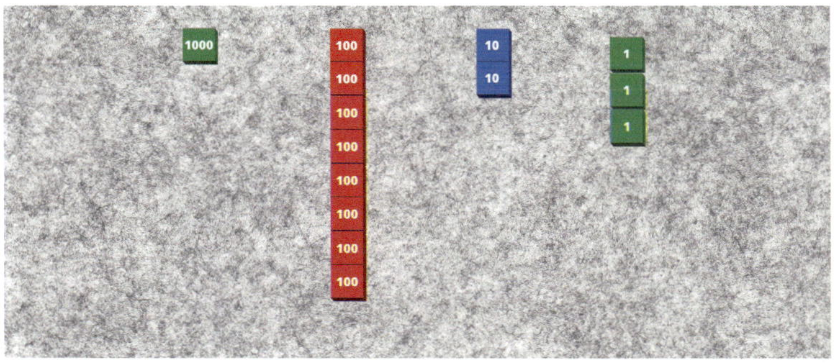


6134

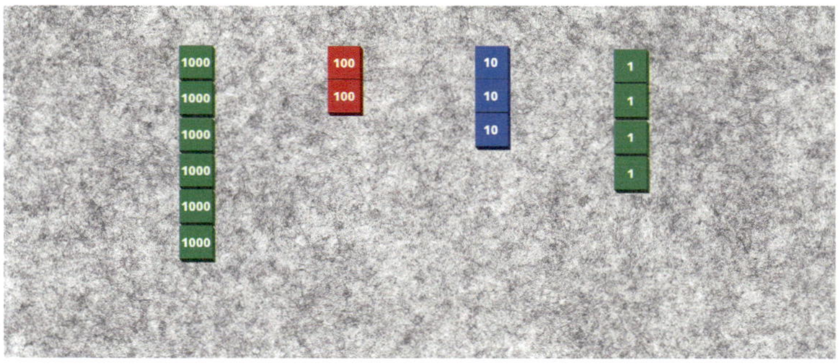

# 5.4 | Stamp Game: Formation of Stamp Quantities

**Materials:**

- Green, red, and blue crayons or pencils

**Directions:**

Invite your student to color in the corresponding stamps for the given numeral. Thousands and units should be green, hundreds should be red, and tens should be blue.

6953

| | | | |
|---|---|---|---|
| 1000 | 100 | 10 | 1 |
| 1000 | 100 | 10 | 1 |
| 1000 | 100 | 10 | 1 |
| 1000 | 100 | 10 | 1 |
| 1000 | 100 | 10 | 1 |
| 1000 | 100 | 10 | 1 |
| 1000 | 100 | 10 | 1 |
| 1000 | 100 | 10 | 1 |
| 1000 | 100 | 10 | 1 |

4571

| | | | |
|---|---|---|---|
| 1000 | 100 | 10 | 1 |
| 1000 | 100 | 10 | 1 |
| 1000 | 100 | 10 | 1 |
| 1000 | 100 | 10 | 1 |
| 1000 | 100 | 10 | 1 |
| 1000 | 100 | 10 | 1 |
| 1000 | 100 | 10 | 1 |
| 1000 | 100 | 10 | 1 |
| 1000 | 100 | 10 | 1 |

# 5.5 | Stamp Game Introduction: Part 3

**Materials**:

- Stamp game (skittles)

**Part 3 Directions:**

1. Say, **"Now that you know that these are stamps (point to the stamps), we are going to learn about the skittles (point to the skittles). We use skittles to represent people."**
2. Take out a green skittle. Say, **"This represents one person."** Put it away.
3. Take out a blue skittle. Say, **"This represents ten people."**
4. Take out ten green skittles and put them in one hand. Hold the one blue skittle in your other hand. Say, **"This one blue skittle** (hold up blue skittle) **represents ten people** (hold up the other hand with ten green skittles)." Put skittles away.
5. Take out one red skittle. Say, **"This represents one hundred people! If I wanted to show 100 people, how many green skittles would I need to hold in my hand?"** (Answer: 100!) **"We don't need to do that. We don't have that many green skittles anyway! I am so glad we don't have to count out 100 people. We can just use this one red skittle to represent 100 people!"**
6. Take out the single large green skittle. Say, **"This represents one thousand people!" Ask, "How many little green skittles would I need to hold up for them to equal this one big skittle?"** (Answer: 1000!)
7. Invite your student to complete the next activity as a follow-up to this introduction.

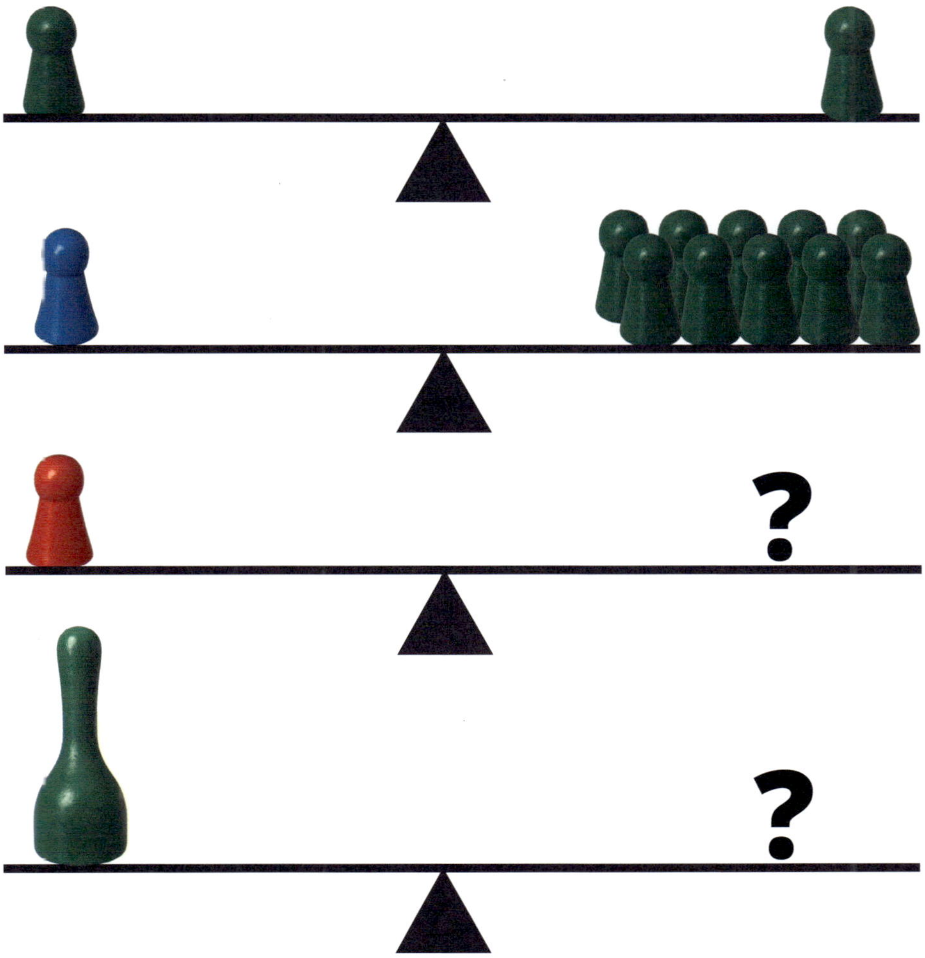

# 5.6 | Stamp Game Introduction: Part 3 Activity

**Directions:**

1. Invite your student to cut out the *skittles* from the bottom of this page and then paste them into the appropriate circle to match the number of people in the image. (Encourage the student to count by tens, then hundreds, to get to 100 then 1000.)
2. Write the number of people represented on the line.

# 5.7 | Stamp Game Introduction: Part 4

**Materials:**

- Stamp Game (stamps and discs)

**Part 4 Directions:**

1. Say, "We have learned that the stamps (point to the stamps) represent amounts, the skittles (point to the skittles) represent people, and now we are going to learn about these discs (point to the discs.)"
2. Say, "These discs look like a number. What number do you think this disc looks like?" Allow time to answer. "Yes, this looks like the number zero! A disc represents the number zero."
3. Say, "We place the disc in any column that has zero stamps, following the leading number. Let's try a couple of problems!"

- [ ] 1. Place the stamps for the number three thousand on the mat. Then, add discs if any columns have zero stamps. Check your answer on the next page.

- [ ] 2. Place the stamps for the number two hundred one on the mat. Then, add discs if any columns after the leading numeral have zero stamps. Check your answer on the next page.

- [ ] 3. Place the stamps for the number one thousand thirty five. Then, add discs if any columns have zero stamps. Check your answer on the next page.

- [ ] 4. Place the stamps for the number four hundred twenty on the mat. Then, add discs if any columns after the leading numeral have zero stamps. Check your answer on the next page.

**Note**: Any columns that come before the *first* written number don't actually need discs. The number is written 201, not 0201. So just make sure you write the number correctly!

1

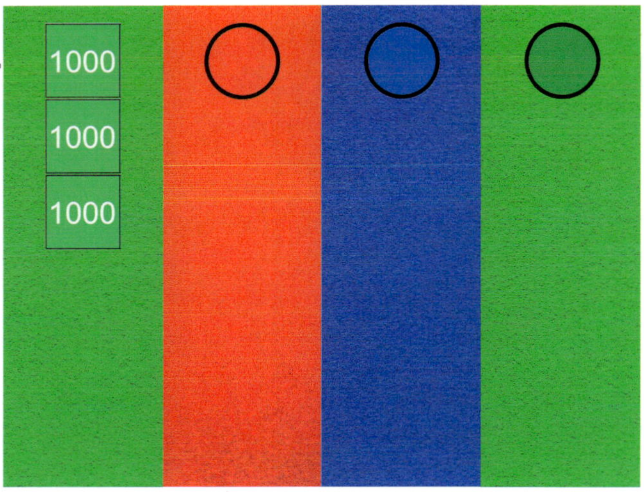


2

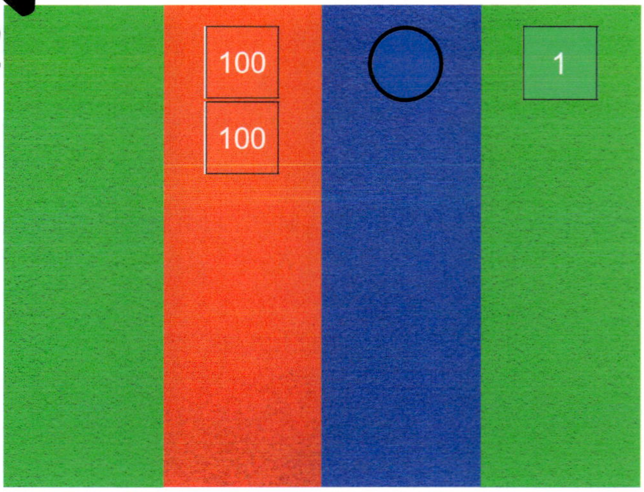


3

4

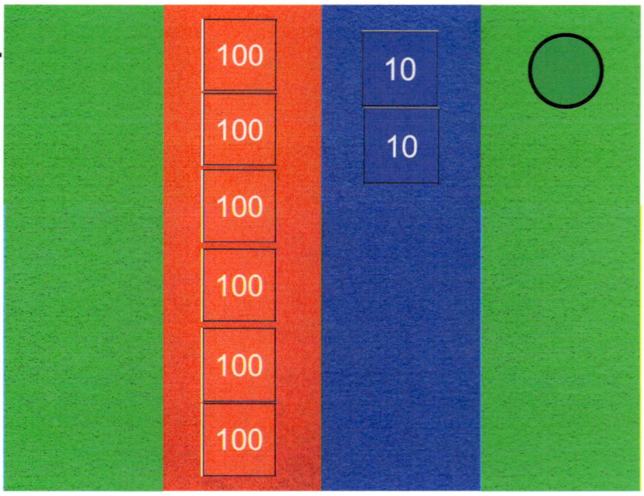

# 5.8 | Stamp Game: Formation of Stamp Quantities with Zeroes

**Materials:** Green, red, and blue crayons or pencils

**Directions:**

- Invite your student to color in the corresponding stamps for the given numeral. Thousands and units should be green, hundreds should be red, and tens should be blue. (NOTE: You may need to invite your child to underline each place value before starting with the proper color of pencil, starting with the units in green and moving left (green, blue, red, green from right to left.)
- For the columns that should be "zero," have your child color in the disc at the top of the column of stamps.

704

| | | | |
|---|---|---|---|
| 1000 | 100 | 10 | 1 |
| 1000 | 100 | 10 | 1 |
| 1000 | 100 | 10 | 1 |
| 1000 | 100 | 10 | 1 |
| 1000 | 100 | 10 | 1 |
| 1000 | 100 | 10 | 1 |
| 1000 | 100 | 10 | 1 |
| 1000 | 100 | 10 | 1 |
| 1000 | 100 | 10 | 1 |

2050

| | | | |
|---|---|---|---|
| 1000 | 100 | 10 | 1 |
| 1000 | 100 | 10 | 1 |
| 1000 | 100 | 10 | 1 |
| 1000 | 100 | 10 | 1 |
| 1000 | 100 | 10 | 1 |
| 1000 | 100 | 10 | 1 |
| 1000 | 100 | 10 | 1 |
| 1000 | 100 | 10 | 1 |
| 1000 | 100 | 10 | 1 |

1360

| | | | |
|---|---|---|---|
| 1000 | 100 | 10 | 1 |
| 1000 | 100 | 10 | 1 |
| 1000 | 100 | 10 | 1 |
| 1000 | 100 | 10 | 1 |
| 1000 | 100 | 10 | 1 |
| 1000 | 100 | 10 | 1 |
| 1000 | 100 | 10 | 1 |
| 1000 | 100 | 10 | 1 |
| 1000 | 100 | 10 | 1 |

4016

| | | | |
|---|---|---|---|
| 1000 | 100 | 10 | 1 |
| 1000 | 100 | 10 | 1 |
| 1000 | 100 | 10 | 1 |
| 1000 | 100 | 10 | 1 |
| 1000 | 100 | 10 | 1 |
| 1000 | 100 | 10 | 1 |
| 1000 | 100 | 10 | 1 |
| 1000 | 100 | 10 | 1 |
| 1000 | 100 | 10 | 1 |

9803

| | | | |
|---|---|---|---|
| 1000 | 100 | 10 | 1 |
| 1000 | 100 | 10 | 1 |
| 1000 | 100 | 10 | 1 |
| 1000 | 100 | 10 | 1 |
| 1000 | 100 | 10 | 1 |
| 1000 | 100 | 10 | 1 |
| 1000 | 100 | 10 | 1 |
| 1000 | 100 | 10 | 1 |
| 1000 | 100 | 10 | 1 |

3021

| | | | |
|---|---|---|---|
| 1000 | 100 | 10 | 1 |
| 1000 | 100 | 10 | 1 |
| 1000 | 100 | 10 | 1 |
| 1000 | 100 | 10 | 1 |
| 1000 | 100 | 10 | 1 |
| 1000 | 100 | 10 | 1 |
| 1000 | 100 | 10 | 1 |
| 1000 | 100 | 10 | 1 |
| 1000 | 100 | 10 | 1 |

105

| | | | |
|---|---|---|---|
| 1000 | 100 | 10 | 1 |
| 1000 | 100 | 10 | 1 |
| 1000 | 100 | 10 | 1 |
| 1000 | 100 | 10 | 1 |
| 1000 | 100 | 10 | 1 |
| 1000 | 100 | 10 | 1 |
| 1000 | 100 | 10 | 1 |
| 1000 | 100 | 10 | 1 |
| 1000 | 100 | 10 | 1 |

1005

| | | | |
|---|---|---|---|
| 1000 | 100 | 10 | 1 |
| 1000 | 100 | 10 | 1 |
| 1000 | 100 | 10 | 1 |
| 1000 | 100 | 10 | 1 |
| 1000 | 100 | 10 | 1 |
| 1000 | 100 | 10 | 1 |
| 1000 | 100 | 10 | 1 |
| 1000 | 100 | 10 | 1 |
| 1000 | 100 | 10 | 1 |

# 6.1 | Stamp Game: Addition Instructions

**Materials:**

- Stamp Game
- Green, red, and blue colored pencils

**Background for the Teacher:**

You will demonstrate two addition problems for your student today. The first problem is a static addition problem that will not have exchanges. The second problem will have exchanges. Most problems for use with the Stamp Game should have exchanges, as becoming proficient at exchanging is one of the major goals of the material. (Note: You will not use the skittles in these addition problems.)

**Directions for Problem #1:**

1. Retrieve the stamps for the first addend, lining them up vertically in groups.

2. Place a pencil underneath the first addend to separate it from the second addend.

3. Underneath the pencil, lay out the stamps for your second addend.

4. Start adding with the units. Remove the pencil. Push the units together then count them by moving them slightly and re-aligning them as you count. (This gives the student the tactile experience of counting and prevents many errors!)

5. Next, add the tens, then the hundreds, then the thousands.

6. Lastly, record the answer in the grid using the proper colored pencils. (You will get: 7,787.)

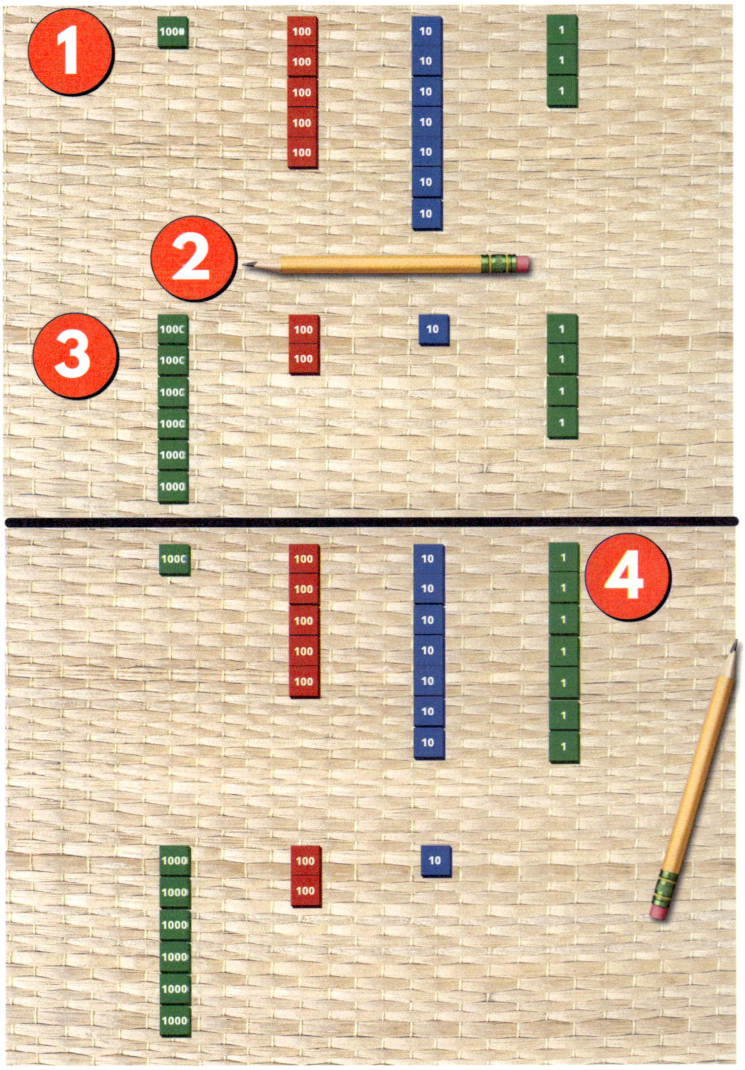

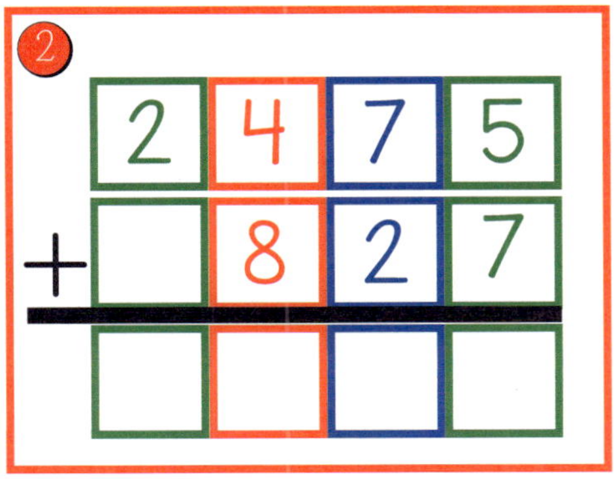


**Directions for Problem #2:**

1. Retrieve the stamps for the first addend, lining them up vertically in groups.

2. Place a pencil underneath the first addend to separate it from the second addend.

3. Underneath the pencil, lay out the stamps for your second addend.

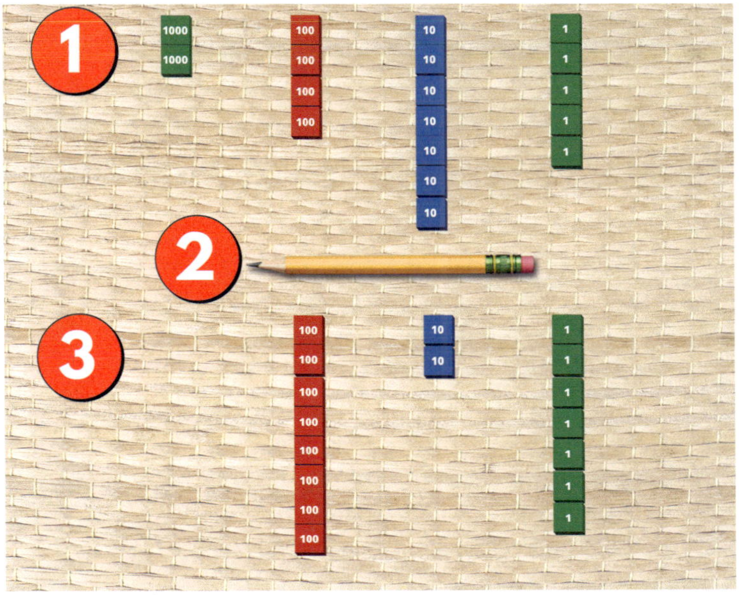


4. Start adding with the units. Remove the pencil. Push the units together then count them by moving them slightly and re-aligning them as you count. (This gives the student the tactile experience of counting and prevents many errors!) Stop when you get to ten! (**See Step 5 on next page.)**

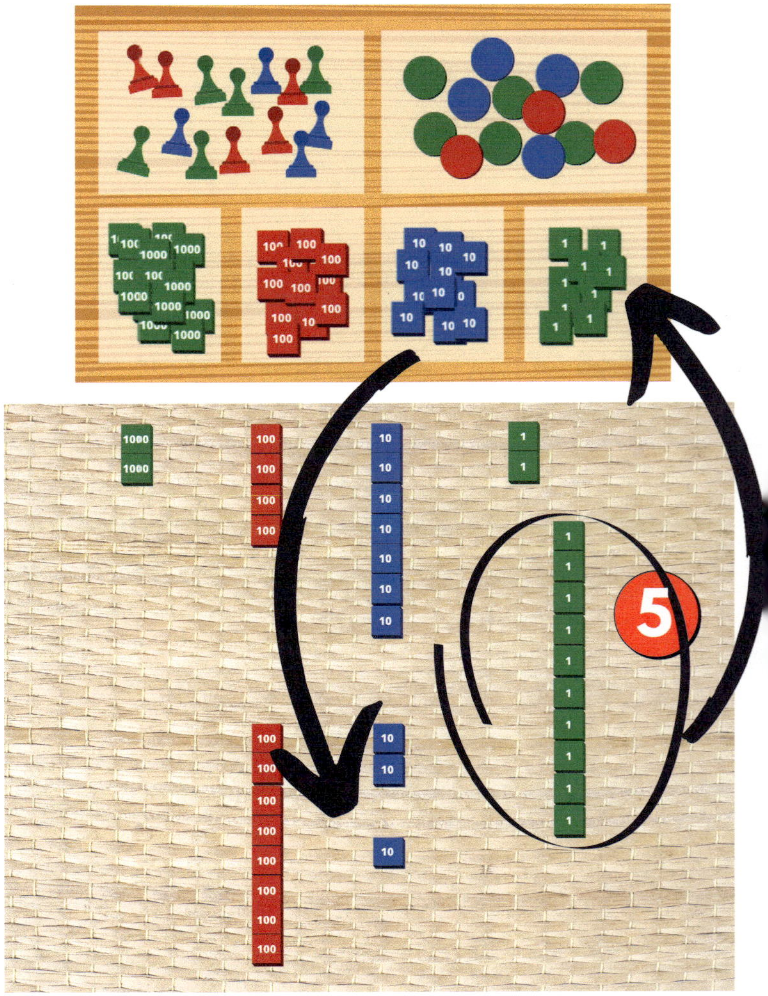

5. Say to your student, "I have ten units. I need to exchange these ten units for one ten." Remove the ten units that you have just counted and aligned off to the side and place them in the Stamp Game box.

Remove one blue 10 stamp and place it underneath the column of blue ten stamps, leaving a small space below the stamps of the second addend.

6. Finish counting the remaining units. (There are two.)

7. Push the tens together then count. There are ten, so exchange them for a red 100 stamp and place it below the column of red hundred stamps, leaving a small space again. There are no stamps left in the tens column, so place a blue disc in the tens column to represent zero.

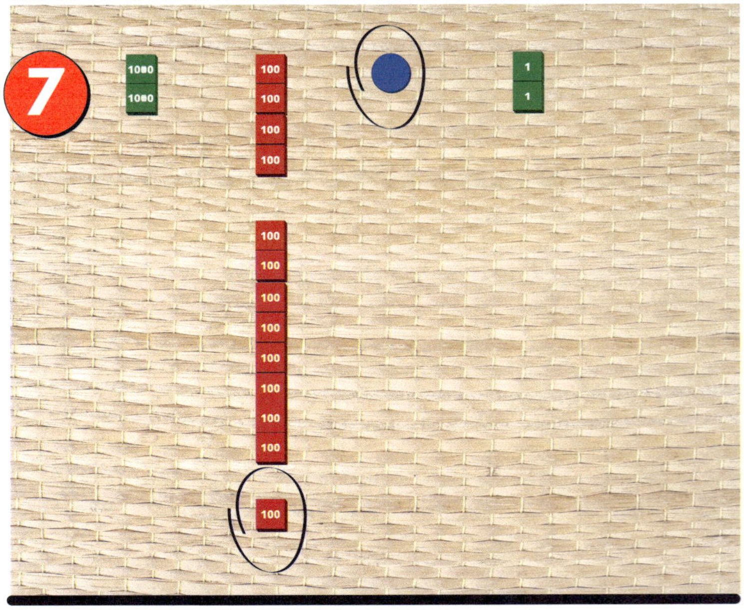

8. Push the hundreds together and count them by moving them slightly to the right in a new column as you count. When you get to ten, exchange the ten 100 stamps for one 1000 stamp.

Place the thousand stamp underneath the existing thousand stamps, leaving a space.

Finish counting the hundreds remaining. (There are 3.)

9. Push the thousands together. Count as before.

10 .Record your answer using red, blue, and green colored pencils. Emphasize the zero in the tens place. (Answer: 3,302.)

# 6.2 | Stamp Game: Addition Problems

**Materials:**

- Stamp Game
- Green, red, and blue colored pencils

**Directions:** Use the Stamp Game to solve. Expect to spend two or more days to work through these 12 problems.

1

| | | | | |
|---|---|---|---|---|
| | 1 | 5 | 7 | 3 |
| + | 6 | 2 | 1 | 4 |
| | | | | |

2

| | | | | |
|---|---|---|---|---|
| | 2 | 4 | 7 | 5 |
| + | | 8 | 2 | 7 |
| | | | | |

3

| | | | | |
|---|---|---|---|---|
| | 8 | 0 | 2 | 2 |
| + | | 7 | 0 | 8 |
| | | | | |

4

| | | | | |
|---|---|---|---|---|
| | 6 | 3 | 5 | 2 |
| + | 3 | 2 | 5 | 1 |
| | | | | |

5

| | | | | |
|---|---|---|---|---|
| | | 2 | 0 | 6 |
| + | | 8 | 2 | 2 |
| | | | | |

6

| | | | | |
|---|---|---|---|---|
| | | | 5 | 8 |
| + | | | 9 | 4 |
| | | | | |

7

| | | | | |
|---|---|---|---|---|
| | 2 | 5 | 7 | 1 |
| + | | 2 | 7 | 9 |
| | | | | |

8

| | | | | |
|---|---|---|---|---|
| | 1 | 2 | 9 | 8 |
| + | | 5 | 8 | 4 |
| | | | | |

9

| | | | | |
|---|---|---|---|---|
| | 4 | 3 | 5 | 3 |
| + | | 5 | 6 | 8 |
| | | | | |

10

| | | | | |
|---|---|---|---|---|
| | | 3 | 8 | 7 |
| + | | 2 | 8 | 4 |
| | | | | |

11

| | | | | |
|---|---|---|---|---|
| | | | 9 | 1 |
| + | | | 4 | 4 |
| | | | | |

12

| | | | | |
|---|---|---|---|---|
| | | | 7 | 8 |
| + | | | 6 | 5 |
| | | | | |

# 6.3 | Stamp Game: Addition Story Problems

**Directions:** Use the stamp game to solve. Color the pictures after checking your work with your teacher.

Sal picked 163 blueberries. Her mother wants her to pick 37 more blueberries to have enough for the pie. How many total blueberries are needed for the pie?

| | | | | |
|---|---|---|---|---|
| | | 1 | 6 | 3 |
| + | | | 3 | 7 |
| | | | | |

Joey is planting apple seeds. He has already planted 86 seeds. He has 25 more seeds to plant. How many seeds will he plant in all?

| | | | | |
|---|---|---|---|---|
| | | | 8 | 6 |
| + | | | | |
| | | | | |

Sebastian has 116 mangoes from his tree to sell at the market stand. Sarah has 59 mangoes from her tree to sell at the market stand. How many mangoes do they have altogether?

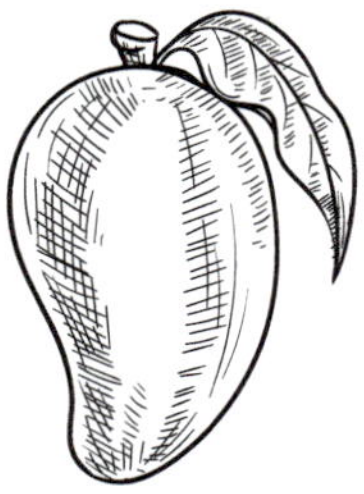

| | | | | |
|---|---|---|---|---|
| | | | | |
| + | | | | |
| | | | | |

# 7.1 | Stamp Game: Subtraction Instructions

**Materials:**
- Stamp Game
- Green, red, and blue colored pencils

**Background for the Teacher:**
You will demonstrate two subtraction problems for your student today. The first problem is a static subtraction problem that will not have exchanges. The second problem will have exchanges. Most problems for use with the Stamp Game should have exchanges, as becoming proficient at exchanging is one of the major goals of the material. (Note: You will not use the skittles in these subtraction problems.)

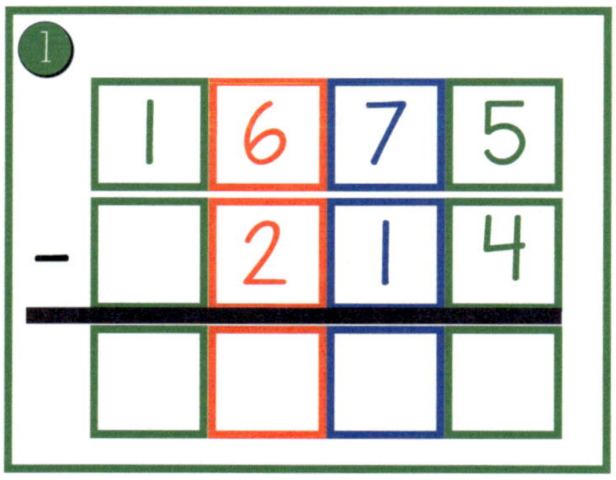

**Directions for Problem #1:**

1. Retrieve the stamps for the minuend, lining them up vertically in groups.

2. Subtract the subtrahend, starting with the units. Remind your student that we only use the bank of stamps for exchanging now. We have to take away the subtrahend from the minuend, and these are all the stamps we have. So, starting with the units, count four units and stack them in a tower. Keep them in the units column, leaving a space. (Mark the space with a pencil.)
Continue with the tens, hundreds, and thousands.

3. The answer will be the **remainder** that is left of the minuend, *above the pencil*. (Circled.)

4. To check your work, count the subtrahend that was subtracted and make sure it matches the written subtrahend in the problem.

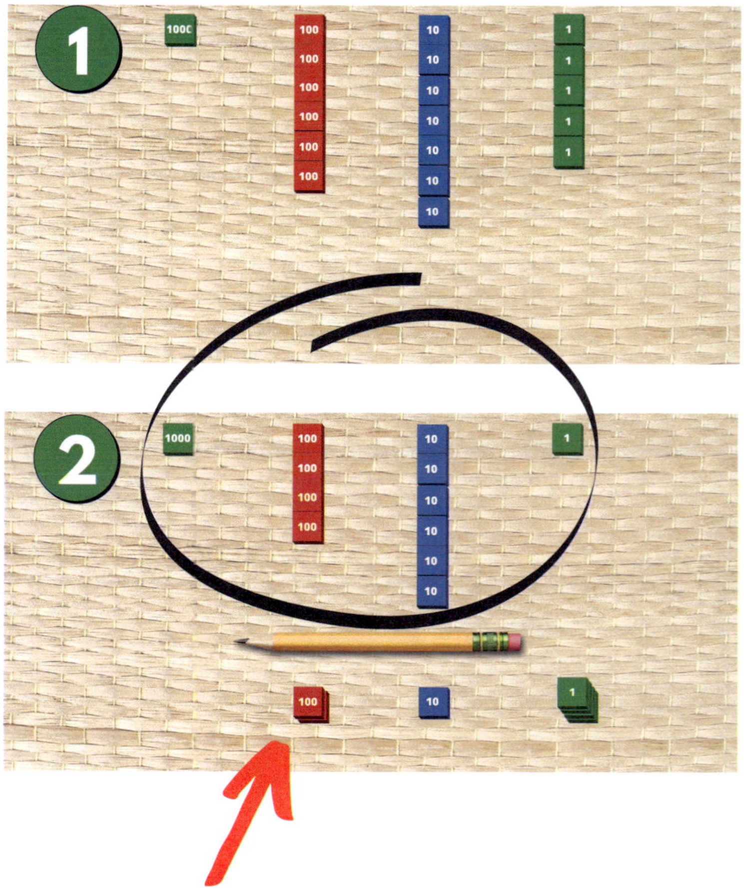


*Stack the stamps of the subtrahend as you subtract.

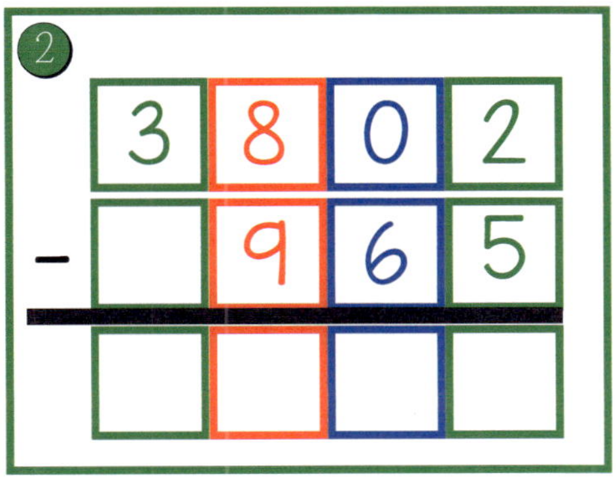

### Directions for Problem #2:

**1.** Retrieve the stamps for the minuend, lining them up vertically in groups.

**2a.** Subtract the subtrahend, starting with the units. Remind your student about using the stamps in the box. **"We only use the bank of stamps for *exchanging* now. We have to take away the subtrahend from the minuend, and these are all the stamps we have."** So, starting with the units, start to count five units and stack them in a tower. You will only be able to count *two* units before having to make an exchange to free up more units!

This is a particularly advanced problem, because you will notice there are no tens in the tens column. Ask your student, **"What do we do if we have no tens here? Do you know where we can get some tens?"**

You will have to go to the hundreds column and exchange one 100 for ten 10's. Place the ten 10's in the tens column first.

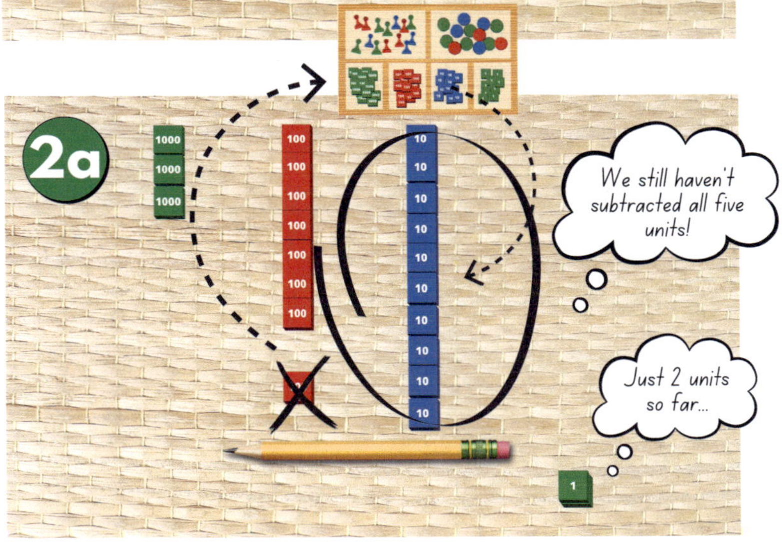


**2b.** Now exchange a single 10 for ten unit stamps.

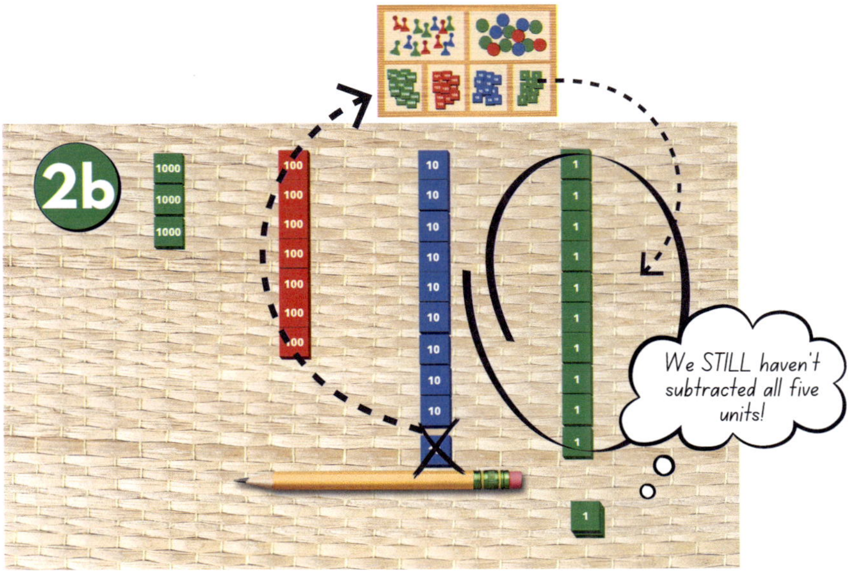

**2c.** Finally, you can continue to subtract the unit stamps to make a total of five units. Stack them in the units column, leaving a space. (Mark the space with a pencil.)

(NOTE: You may need to encourage your student to re-count the stacked units at this point, since two exchanges had to take place to get here!)

**3.** Continue with the tens and hundreds.

Subtract the six 10's and stack them in the tens column.

When you try to subtract nine 100's, you will need to make an exchange. Exchange one 1000 stamp for ten 100 stamps and finish subtracting the hundreds.

**4.** The answer will be the **remainder** that is left of the minuend, *above the pencil*. Invite your student to record the answer in the proper colors of pencil.

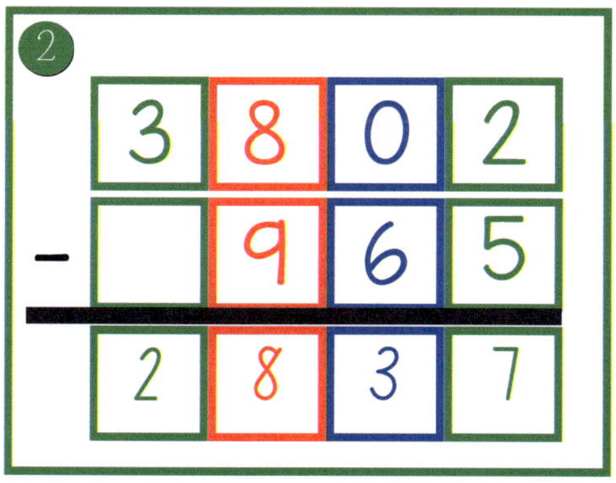

**5.** To check your work, count the subtrahend that was subtracted and make sure it matches the written subtrahend in the problem.

# 7.2 | Stamp Game: Subtraction Problems

**Materials:**

- Stamp Game
- Green, red, and blue colored pencils

**Directions:** Use the Stamp Game to solve. Expect to spend two or more days to work through these 12 problems.

1

| | | | | |
|---|---|---|---|---|
| | 1 | 6 | 7 | 7 |
| − | | 2 | 1 | 4 |
| | | | | |

2

| | | | | |
|---|---|---|---|---|
| | 3 | 8 | 0 | 2 |
| − | | 9 | 6 | 5 |
| | | | | |

3

| | | | | |
|---|---|---|---|---|
| | 4 | 1 | 2 | 3 |
| − | 2 | 3 | 8 | 3 |
| | | | | |

4

| | | | | |
|---|---|---|---|---|
| | 1 | 4 | 1 | 3 |
| − | | 4 | 6 | 2 |
| | | | | |

5

| | | | | |
|---|---|---|---|---|
| | | 1 | 3 | 8 |
| − | | | 8 | 5 |
| | | | | |

6

| | | | | |
|---|---|---|---|---|
| | 4 | 6 | 7 | 7 |
| − | 1 | 3 | 9 | 4 |
| | | | | |

7

| | | | | |
|---|---|---|---|---|
| | | 9 | 6 | 0 |
| − | | 1 | 4 | 7 |
| | | | | |

8

| | | | | |
|---|---|---|---|---|
| | | 5 | 0 | 6 |
| − | | 1 | 8 | 5 |
| | | | | |

9

| | | | | |
|---|---|---|---|---|
| | 8 | 0 | 3 | 8 |
| − | 2 | 9 | 2 | 6 |
| | | | | |

10

| | | | | |
|---|---|---|---|---|
| | 3 | 2 | 2 | 1 |
| − | | 6 | 7 | 5 |
| | | | | |

11

| | | | | |
|---|---|---|---|---|
| | | | 8 | 6 |
| − | | | 3 | 7 |
| | | | | |

12

| | | | | |
|---|---|---|---|---|
| | | 4 | 7 | 9 |
| − | | 1 | 8 | 1 |
| | | | | |

# 7.3 | Stamp Game: Subtraction Story Problems

**Directions:** Use the stamp game to solve. Check your work with your teacher and then color the picture.

Mary needs to count all of her sheep. She has 200 sheep in the pasture. So far, she has counted 87 sheep. How many more sheep does she need to count?

| | | | | |
|---|---|---|---|---|
| | | 2 | 0 | 0 |
| − | | | 8 | 7 |
| | | | | |

Farmer Abe has 102 horses. He plans to sell 28 horses at the horse auction. If he sells all 28 horses, how many will he have left?

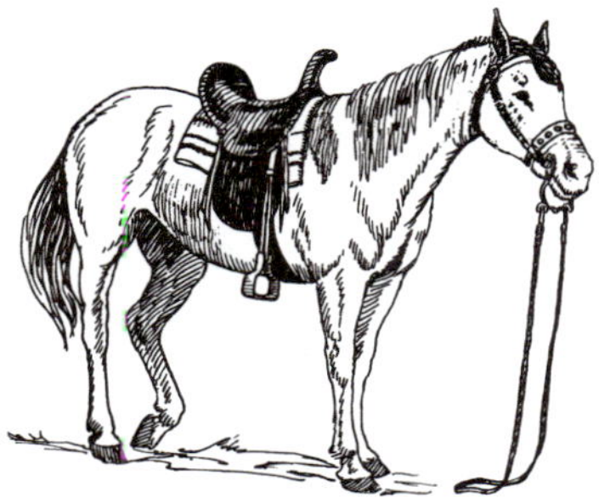

| | | | | |
|---|---|---|---|---|
| | | 1 | 0 | 2 |
| − | | | | |
| | | | | |

Isabella has 64 eggs to sell at the market. She sells 36 eggs to the first family to visit her stand! How many eggs does she have left to sell?

| | | | | |
|---|---|---|---|---|
| | | | | |
| − | | | | |
| | | | | |

# 8.1 | Stamp Game: Multiplication Instructions

**Materials:**

- Stamp Game
- Green, red, and blue colored pencils

**Background for the Teacher:**

You will demonstrate two multiplication problems for your student today. The first problem is a static multiplication problem that will not have exchanges. The second problem will have exchanges. Most problems for use with the Stamp Game should have exchanges, as becoming proficient at exchanging is one of the major goals of the material.

(Note: You will finally use the skittles in these multiplication problems! Also, you will need an ample amount of vertical space to do these problems, so if you are using a mat, turn it with the short side to the top.)

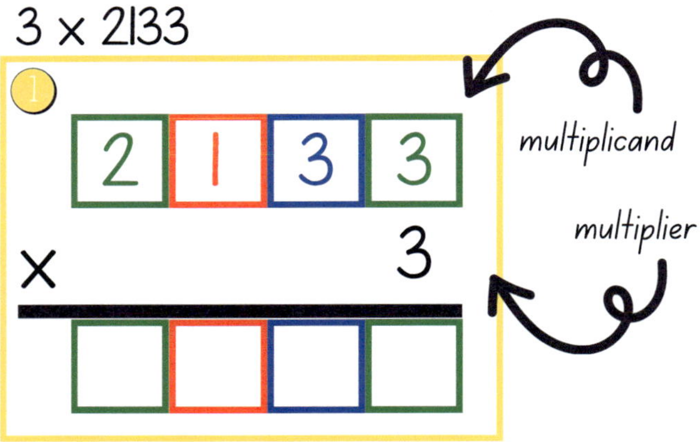

**Directions for Problem #1:**

1. Say, **"We need to multiply this multiplicand (2133) *three times*. We are finally going to use the Skittles to represent the multiplier of 3."** Point to the multiplicand and multiplier in the horizontal equation above the problem box and inside the problem box.

2. Retrieve 3 small green skittles and lay them in a vertical column on the left side of the mat.

3. Retrieve the stamps for the multiplicand, lining them up vertically in a row. Do this *three times*, leaving a space between groups.

4. Say, **"Now we combine these groups by adding them together."** Push the rows of tiles together into lcolumns.

5. Count the skittles in each category, **starting with the units**, and make sure to add a disc for zero if there is a need for a placeholder. (Note: There is no zero category in this particular problem.)

6. Record the answer on the grid inside the problem box. (Not shown. ANSWER: 6399.)

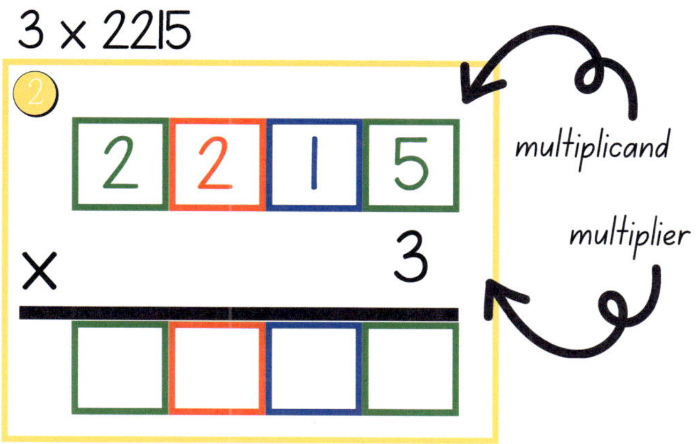

**Directions for Problem #1:**

1. Say, **"We need to multiply this multiplicand (2215) *three times*. We are finally going to use the Skittles to represent the multiplier of 3."** Point to the multiplicand and multiplier in the horizontal equation above the problem box and inside the problem box.

2. Retrieve 3 small green skittles and lay them in a vertical column on the left side of the mat.

3. Retrieve the stamps for the multiplicand, lining them up vertically in the first row, as shown. Do this *three times*, leaving a space between groups.

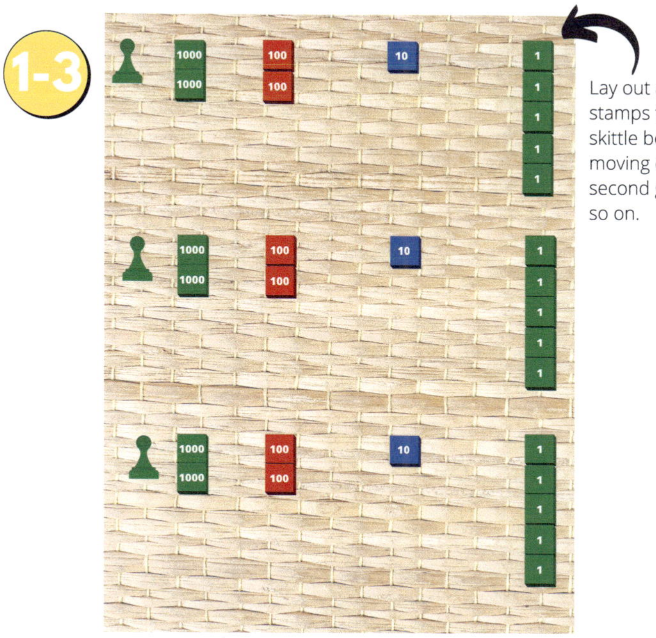


Lay out all of the stamps for the first skittle before moving onto the second group, and so on.

4. Say, **"Now we combine these groups by adding them together."** Push the rows of tiles together into Icolumns.

5. Count the skittles in the unit category first. When you get to ten, ask, **"What can I exchange for these ten units?"** (Pause.) **"Yes, one ten!"**

Stack the ten units, place them in the Stamp Game storage container, and take out one ten stamp. Place the ten stamp below the column of tens, leaving a slight space. (See image on next page.)

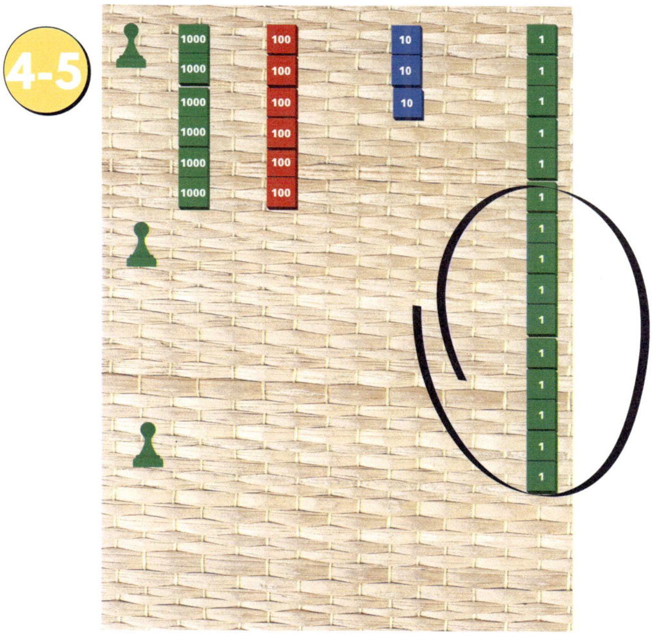

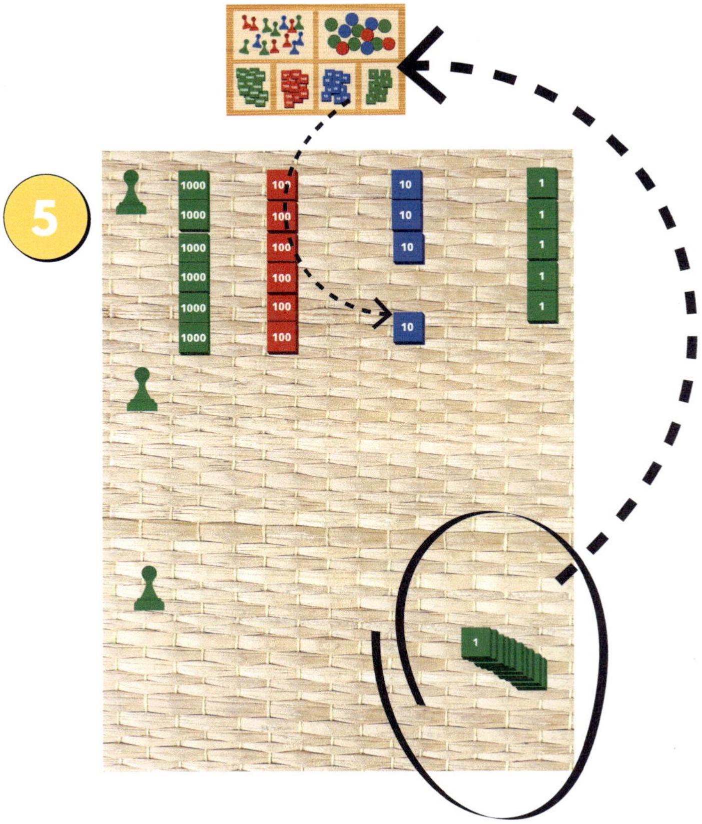


6. Count the remaining units, tens, hundreds, then thousands, making exchanges for ten of any category as needed. (Note: There are no more exchanges in this sample problem.)

7. Record your answer. (Not shown. ANSWER: 6645.)

# 8.2 | Stamp Game: Multiplication Problems

**Materials:**

- Stamp Game
- Green, red, and blue colored pencils

**Directions:** Use the Stamp Game to solve. Expect to spend two or more days to work through these 12 problems.

1

| | 2 | 1 | 3 | 3 |
|---|---|---|---|---|
| × | | | | 3 |
| | | | | |

2

| | 2 | 2 | 1 | 5 |
|---|---|---|---|---|
| × | | | | 3 |
| | | | | |

3

| | 4 | 3 | 2 | 2 |
|---|---|---|---|---|
| × | | | | 2 |
| | | | | |

4

| | 2 | 2 | 4 | 3 |
|---|---|---|---|---|
| × | | | | 3 |
| | | | | |

5

| | 3 | 6 | 1 | 0 |
|---|---|---|---|---|
| × | | | | 2 |
| | | | | |

6

| | 1 | 5 | 2 | 4 |
|---|---|---|---|---|
| × | | | | 3 |
| | | | | |

7

| | | | | |
|---|---|---|---|---|
| | 2 | 4 | 4 | 5 |
| × | | | | 3 |
| | | | | |

8

| | | | | |
|---|---|---|---|---|
| | 2 | 3 | 4 | 4 |
| × | | | | 3 |
| | | | | |

9

| | | | | |
|---|---|---|---|---|
| | 4 | 7 | 5 | 2 |
| × | | | | 2 |
| | | | | |

10

| | | | | |
|---|---|---|---|---|
| | 2 | 1 | 4 | 8 |
| × | | | | 3 |
| | | | | |

11

| | | | | |
|---|---|---|---|---|
| | 3 | 0 | 1 | 0 |
| × | | | | 2 |
| | | | | |

12

| | | | | |
|---|---|---|---|---|
| | | 5 | 2 | 4 |
| × | | | | 3 |
| | | | | |

# 8.3 | Stamp Game: Multiplication Extension 1

**Directions:**
Match the Stamp Game layout to the problem it represents.

Note: The multiplication sentence written horizontally, like 2 x 1213, can be read, "2 groups of 1213."

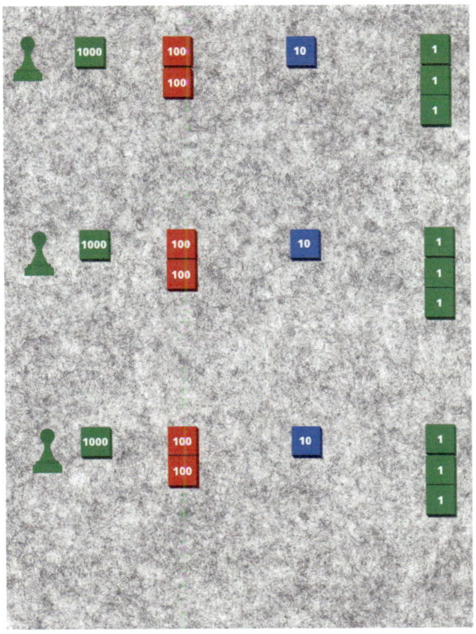


2 x 243

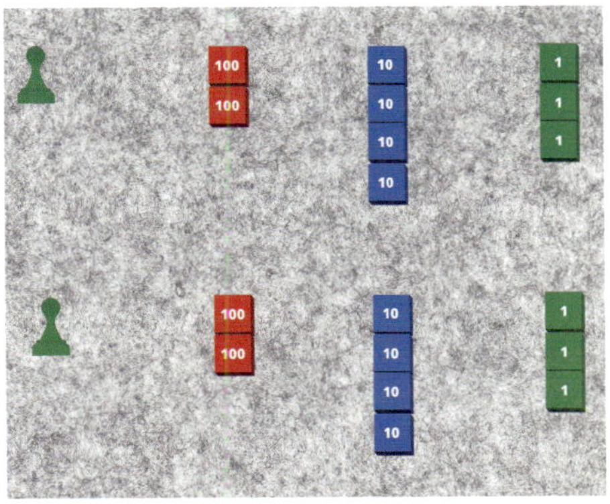


2 x 3215

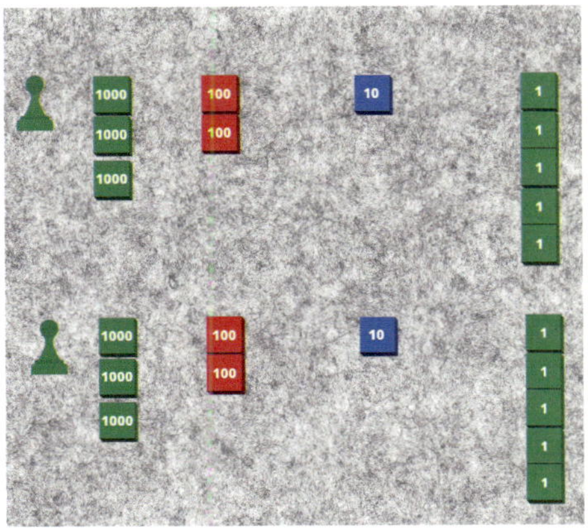


3 x 1213

# 8.4 | Stamp Game: Multiplication Extension 2

**Directions:**
Write both forms of the equation represented on the mat. Solve by counting, crossing out, and drawing new stamps when an exchange is needed OR by using the physical Stamp Game.

1

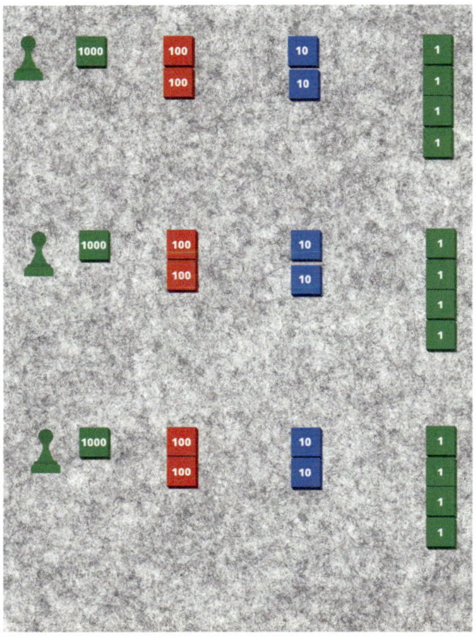


___ × ______ = ______

×

2

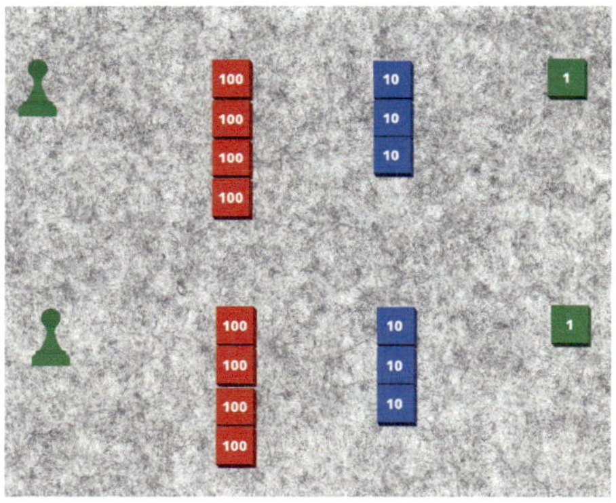


___ × ______ = ______

×

3

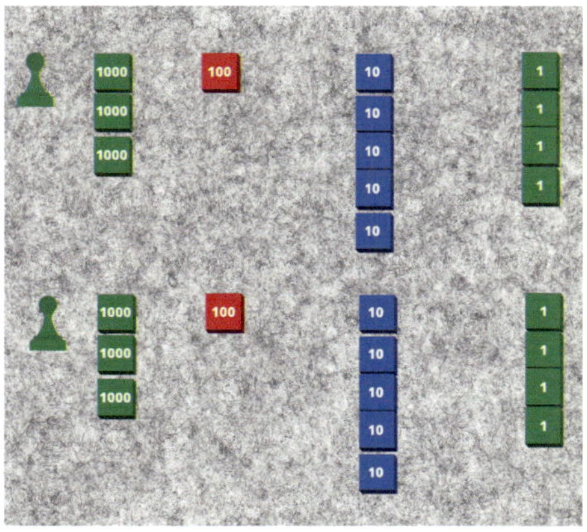


___ × ______ = ______

×

# 8.5 | Stamp Game: Multiplication with 2-Digit Multiplier

**Directions:**
Complete the problems using the green unit skittles. For example, in problem #1, place 11 unit skittles in a column and lay out stamps for the number 231 next to each green skittle. Add them all together at the end, making exchanges as necessary.

1. 11 × 231 = ______

2. 12 × 124 = ______

3. 10 × 311 = ______

4. 13 × 242 = ______

# 8.6 | Stamp Game: Multiplication Story Problems

**Directions:** Use the stamp game to solve. Check your work with your teacher and then color the picture.

Evan is making his own paper muffin cups. Each muffin cup needs to have 44 creases. If he makes 6 muffin cups, how many creases will he make in all?

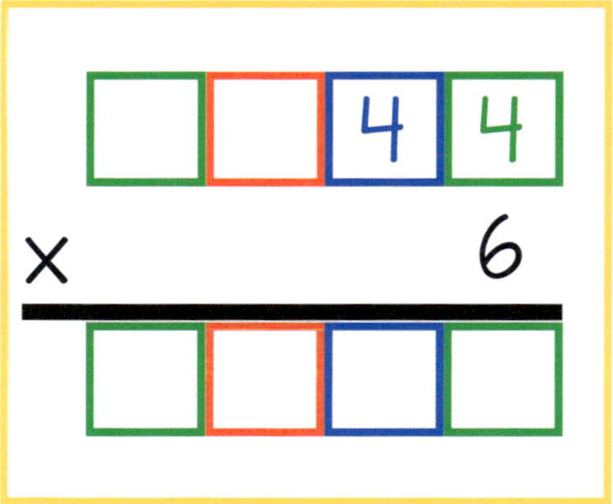


Ellie is in charge of the cookies for the bake sale. She tells 4 of her friends to make a dozen cookies each. She will put all of the cookies into the cookie jar and bring it to the bake sale. How many cookies will be in the jar?
(Note: One dozen is 12 cookies.)

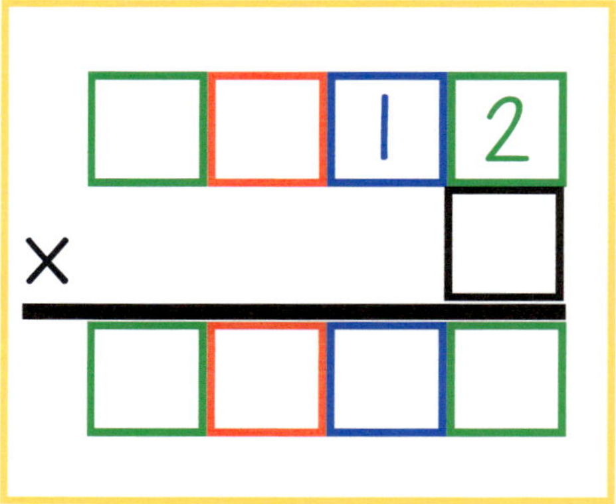


Elmer has a big job. He is in charge ordering the birthday candles for all of the five-year olds in his class. Each five-year old gets 5 candles on his or her cake. This year, there are 6 five-year-olds in his class. How many candles does he need to order?

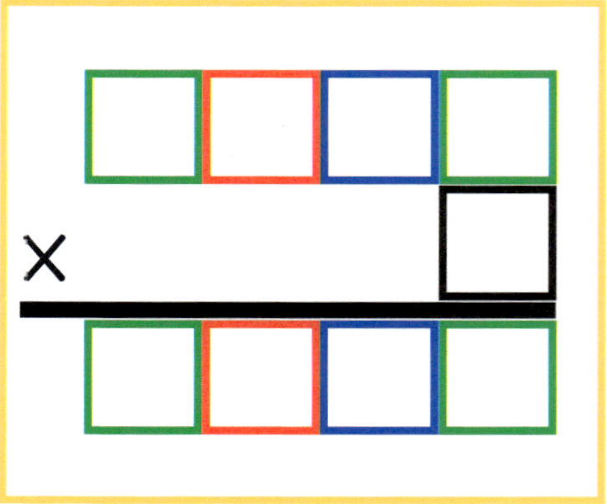

# 9.1 | Stamp Game: Division Instructions

**Materials:**
- Stamp Game
- Green, red, and blue colored pencils

**Background for the Teacher:**
You will demonstrate two division problems for your student today. The first problem is a static division problem that will not have exchanges. The second problem will have exchanges. Most problems for use with the Stamp Game should have exchanges, as becoming proficient at exchanging is one of the major goals of the material.

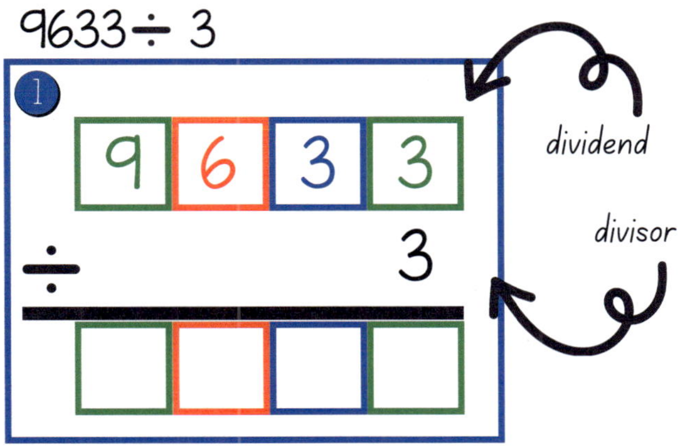

**Directions for Problem #1:**

1. Say, **"We need to divide this dividend, 9633, by this divisor, 3. That means we need to split this number equally between three people."** Make sure to point to dividend and divisor in the problem.

Set out 9633 in stamps as shown. Place three green skittles in a row underneath the stamps as shown.

1

2. Start with the largest unit, which in this case is the thousands. Divide the thousands equally among the three green skittles, giving one stamp at a time.

2

3. Continue giving one thousand stamp per green skittle until they are gone.

3

4. Next, move onto the next unit down - which in this case is the hundreds. Continue giving one stamp to each green skittle in turn until they are gone.

4

5. Continue with the tens, then the units.

6. The answer is the amount **one skittle** receives, which is circled.

7. Record the answer (not shown). ANSWER: 3211

5-6

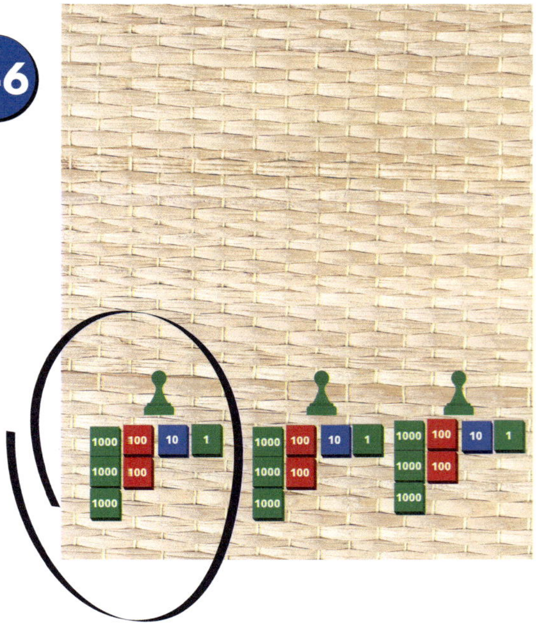

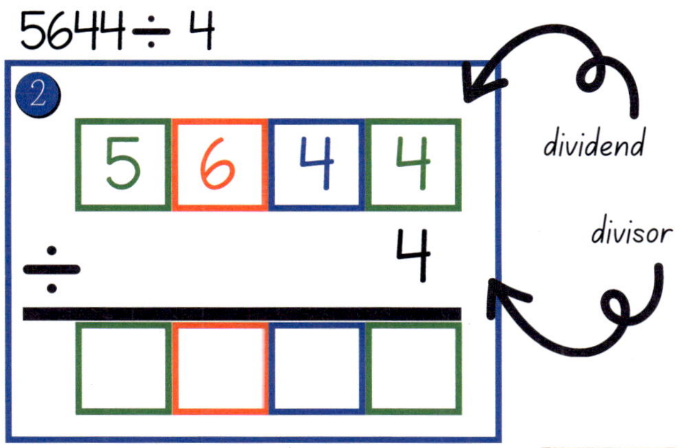

**Directions for Problem #1:**

1. Say, **"We need to divide this dividend, 5644, by this divisor, 4. That means we need to split this number equally between four people."** Make sure to point to dividend and divisor in the problem.

Set out 5644 in stamps as shown. Place four green skittles in a row underneath the stamps as shown.

2. Start with the largest unit, which in this case is the thousands. Divide the thousands equally among the three green skittles, giving one stamp at a time.

Note: There is one leftover thousand stamp that cannot be divided equally among the four green skittles. (See step 3!)

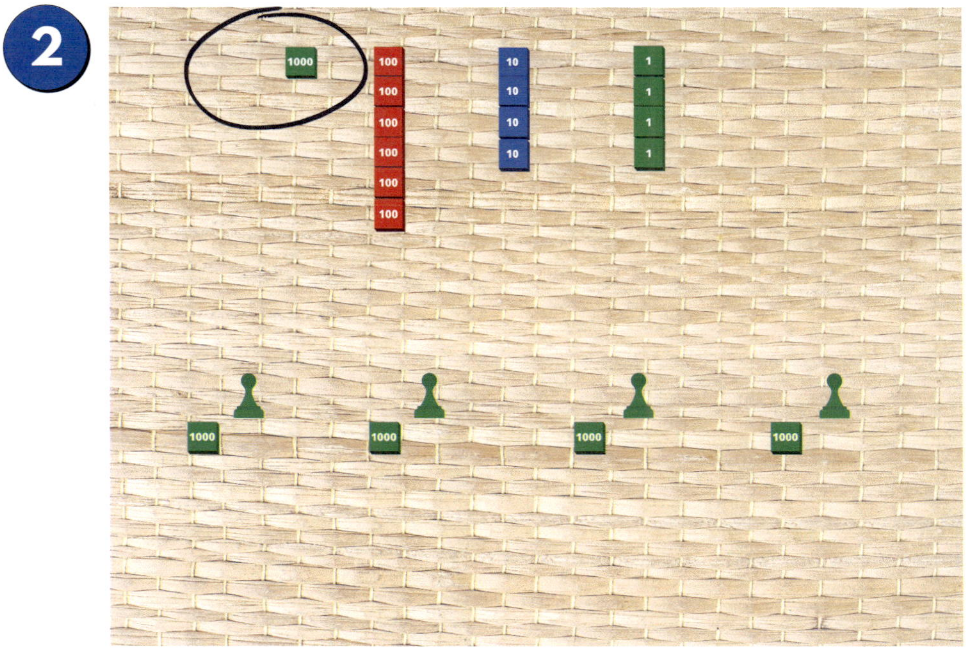

3

3. Since the last thousand stamp cannot be divided equally among the four green skittles, you will need to exchange it for ten of the next lower unit. One thousand stamp gets placed back in the Stamp Game organizer and ten hundred stamps are taken out and placed with the other hundred stamps on your mat.

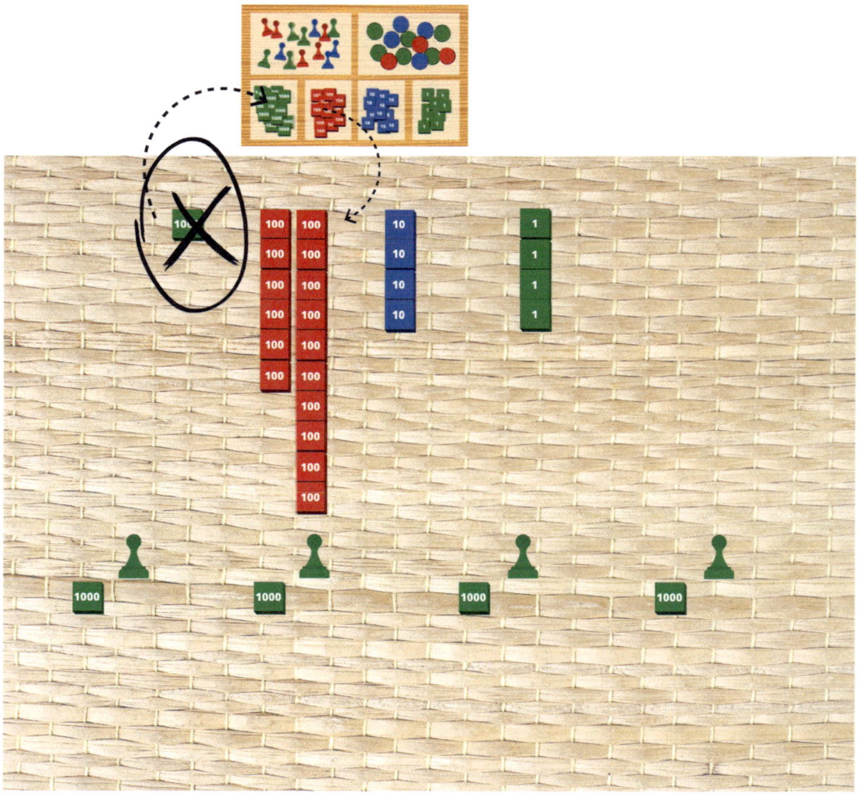

4

4. Now, distribute the hundred stamps equally among the green skittles, one at a time until they are gone.

5. Distribute the ten stamps equally among the green skittles, one at a time.

5-6

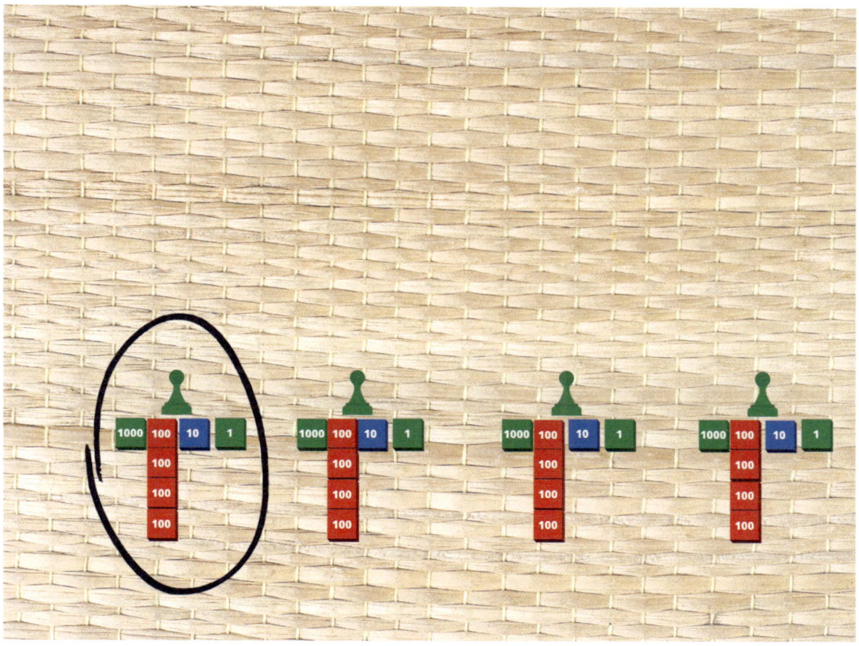


6. Distribute the unit stamps equally among the green skittles, one at a time.
7.The answer is the amount **one skittle** receives. Record the answer. (Not shown. Answer: 1411.)

# 9.2 | Stamp Game: Division Problems

1

| 9 | 6 | 3 | 3 |
|---|---|---|---|

÷ 3

| | | | |
|---|---|---|---|

2

| 5 | 6 | 4 | 4 |
|---|---|---|---|

÷ 4

| | | | |
|---|---|---|---|

3

| 4 | 9 | 2 | 8 |
|---|---|---|---|

÷ 4

| | | | |
|---|---|---|---|

4

| 9 | 6 | 4 | 2 |
|---|---|---|---|

÷ 3

| | | | |
|---|---|---|---|

5

| 5 | 5 | 6 | 0 |
|---|---|---|---|

÷ 5

| | | | |
|---|---|---|---|

6

| 7 | 8 | 4 | 2 |
|---|---|---|---|

÷ 3

| | | | |
|---|---|---|---|

# 9.3 | Stamp Game: Division Problems with Remainders

Sometimes when we divide, we have stamps leftover that cannot be shared equally among the skittles. We call the leftover stamps the "remainder." Each of the problems on this page has a remainder. Write the quotient, then write the remainder in the box.

1. 4873 ÷ 4 = ☐☐☐☐ R = ___

2. 6815 ÷ 3 = ☐☐☐☐ R = ___

3. 4325 ÷ 2 = ☐☐☐☐ R = ___

4. 7558 ÷ 5 = ☐☐☐☐ R = ___

5. 7869 ÷ 6 = ☐☐☐☐ R = ___

6. 5147 ÷ 3 = ☐☐☐☐ R = ___

# 9.4 | Stamp Game: Division with 2-Digit Divisor Instructions

**Materials:**
- Stamp Game
- Green, red, and blue colored pencils

**Background for the Teacher:**
You will demonstrate two division problems for your student today. The first problem is a static division problem with a 2-digit divisor that will not have exchanges. The second problem will have exchanges. Most problems for use with the Stamp Game should have exchanges, as becoming proficient at exchanging is one of the major goals of the material.

Note: You will use the skittles and an ample amount of space today.

1. $2553 \div 23$ (dividend: 2553; divisor: 23)

**Directions for Problem #1:**

**1**. Say, "We need to divide this dividend, 2553, by this divisor, 23. That means we need to split this number equally between twenty-three people." Make sure to point to dividend and divisor in the problem. Set out the 2553 stamps, as shown.

**2.** Say, "Twenty-three is made up of two tens and three units. We will use two blue skittles to represent **twenty people** and three green skittles to represent **three people**." Place two blue skittles and say, "Ten, twenty." Place each green skittle and count by ones up to twenty-three: "twenty-one, twenty-two, twenty-three." Say, "We need to divide these stamps equally among these twenty-three people. Remember, each blue stamp is **ten people** and each green stamp is only **one person**."

**1**

**3.** Start with the largest unit. Say, "First we will distribute the largest unit, which is the thousand unit." Place one thousand stamp underneath the first blue skittle. Then place another thousand stamp underneath the second blue skittle. Ask, "How many people does this blue skittle represent?" [Answer: Ten.] "Since we gave ten people 1000 units to share among them, each person really only got 100. Ten hundreds is the same as one thousand. So, since each person only got one hundred, we need to give each of these green skittles one hundred."

**4.** Distribute one 100 stamp to each green skittle, as shown.

[Note: When you move down one order in the decimal system, the stamp you distribute has to move down one order, as well. So, if you give the blue skittle 1000, you will give the green skittle ten times less, or 100. Your student may or may not pick up on this pattern on her own.]

**3-4**

**5.** Now you have no more thousands. It is time to distribute the hundreds. Give each blue skittle one hundred stamp. Ask, "Since each blue skittle represents ten people, how many did each person get when I gave them each a 100 stamp?" [Answer: 10, since 10 tens is 100.] "Right, each person got ten. So I need to share equally by giving each green skittle a ten stamp."

**5**

**6.** Now you have no more hundreds. It is time to move onto the tens. Give one ten to each of the blue skittles. Ask, "Since each blue skittle represents ten people, how many did each person get when I have them each one ten-stamp?" [Answer: One, since there are ten ones in each ten.] "Yes, one. So each green skittle needs 1 unit stamp to share equally." Distribute the unit stamps.

6

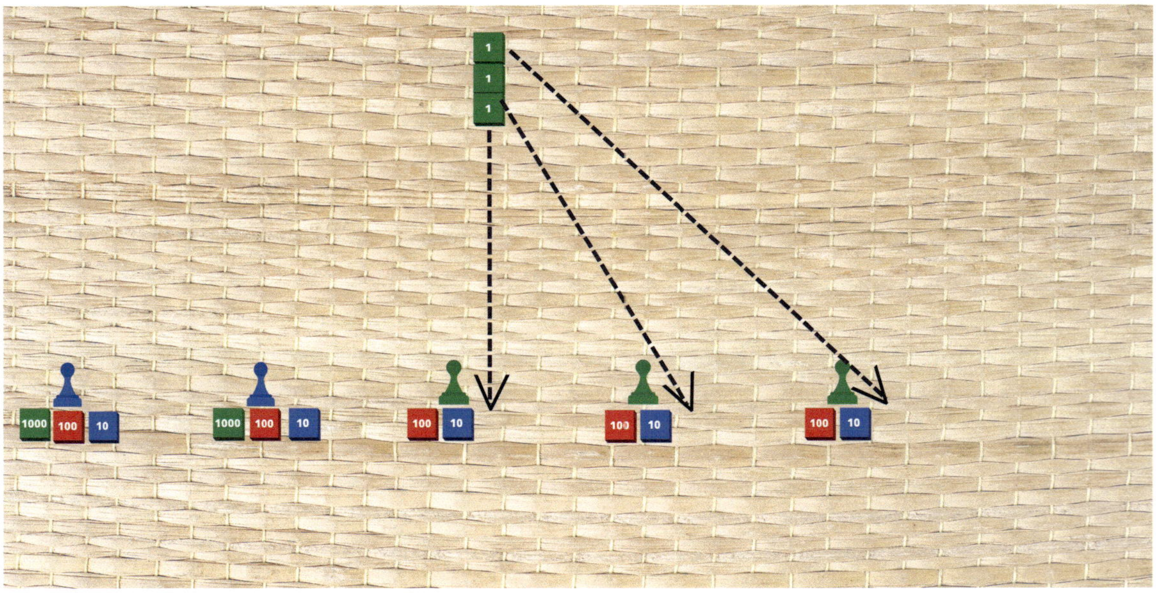


**7.** The answer is the amount that one green skittle receives. [Answer: 111.]

7

# 9.5 | Stamp Game: Division with 2-Digit Divisor Problems (no exchanges)

1. $2553 \div 23 =$ ____________

2. $6752 \div 32 =$ ____________

3. $3926 \div 13 =$ ____________

4. $2814 \div 14 =$ ____________

5. $1919 \div 19 =$ ____________

6. $4662 \div 42 =$ ____________

# 9.6 | Stamp Game: Division with 2-Digit Divisor - Advanced Instructions

**Materials:**

- Stamp Game
- Green, red, and blue colored pencils

**Background for the Teacher:**

You will demonstrate two division problems for your student today. The first problem is a static division problem that will not have exchanges. The second problem will have exchanges. Most problems for use with the Stamp Game should have exchanges, as becoming proficient at exchanging is one of the major goals of the material.

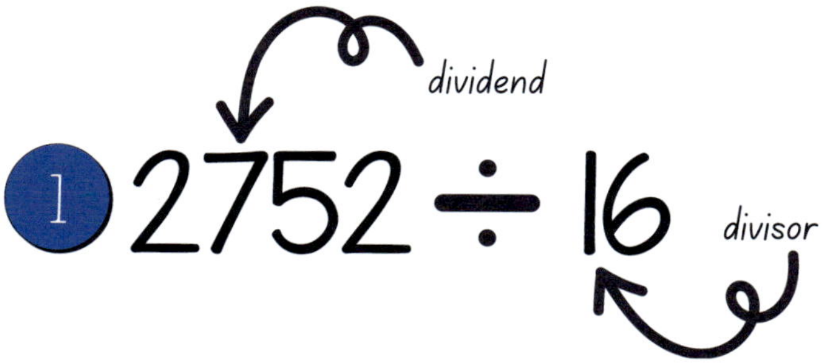

**Directions for Problem #1:**

**1**. Say, "We need to divide this dividend, 2752, by this divisor, 16. That means we need to split this number equally between sixteen people." Make sure to point to dividend and divisor in the problem. Set out the 2752 stamps, as shown.

**2.** Say, "Sixteen is made up of one ten and six units. We will use a blue skittle to represent **ten people** and six green skittles to represent **six people**." Place one blue skittle and say, "Ten." Place each green skittle and count by ones up to sixteen: "eleven, twelve, thirteen, fourteen, fifteen, sixteen." Say, "We need to divide these stamps equally among these sixteen people. Remember, this blue stamp is **ten people** and these green stamps are each only **one person**."

**3.** Start with the largest unit. Say, "First we will distribute the largest unit, which is the thousand unit." Place one thousand stamp underneath the blue skittle. Ask, "How many people does this blue skittle represent?" [Answer: Ten.] "Since we gave ten people 1000 units to share among them, they really each only got 100. Ten hundreds is the same as one thousand. So, since each person only got one hundred, we need to give each of these green skittles one hundred."

**4.** Distribute one 100 stamp to each green skittle, as shown.

[Note: When you move down one order in the decimal system, the stamp you distribute has to move down one order, as well. So, if you give the blue skittle 1000, you will give the green skittle ten times less, or 100. Your student may or may not pick up on this pattern on her own.]

**5.** Say, "Now we are back at the thousands. There is one thousand stamp remaining. If you give the thousand stamp to the blue skittle, there are not enough hundred stamps to distribute to each green stamp, so that's not fair."

"We can exchange this 1000 stamp for ten 100 stamps so we can be fair and give everyone an equal share." See the exchange in the illustration below.

6. Say, "Now let's distribute the 100 stamps. The blue skittle gets one 100. That means each green skittle gets one 10." When you distribute the tens, you need one more for the last skittle. Ask, "How can we get more tens?" [Exchange a 100 stamp! See step 7.]

7.Exchange one 100 stamp for ten 10 stamps then finish distributing to the last green skittle.

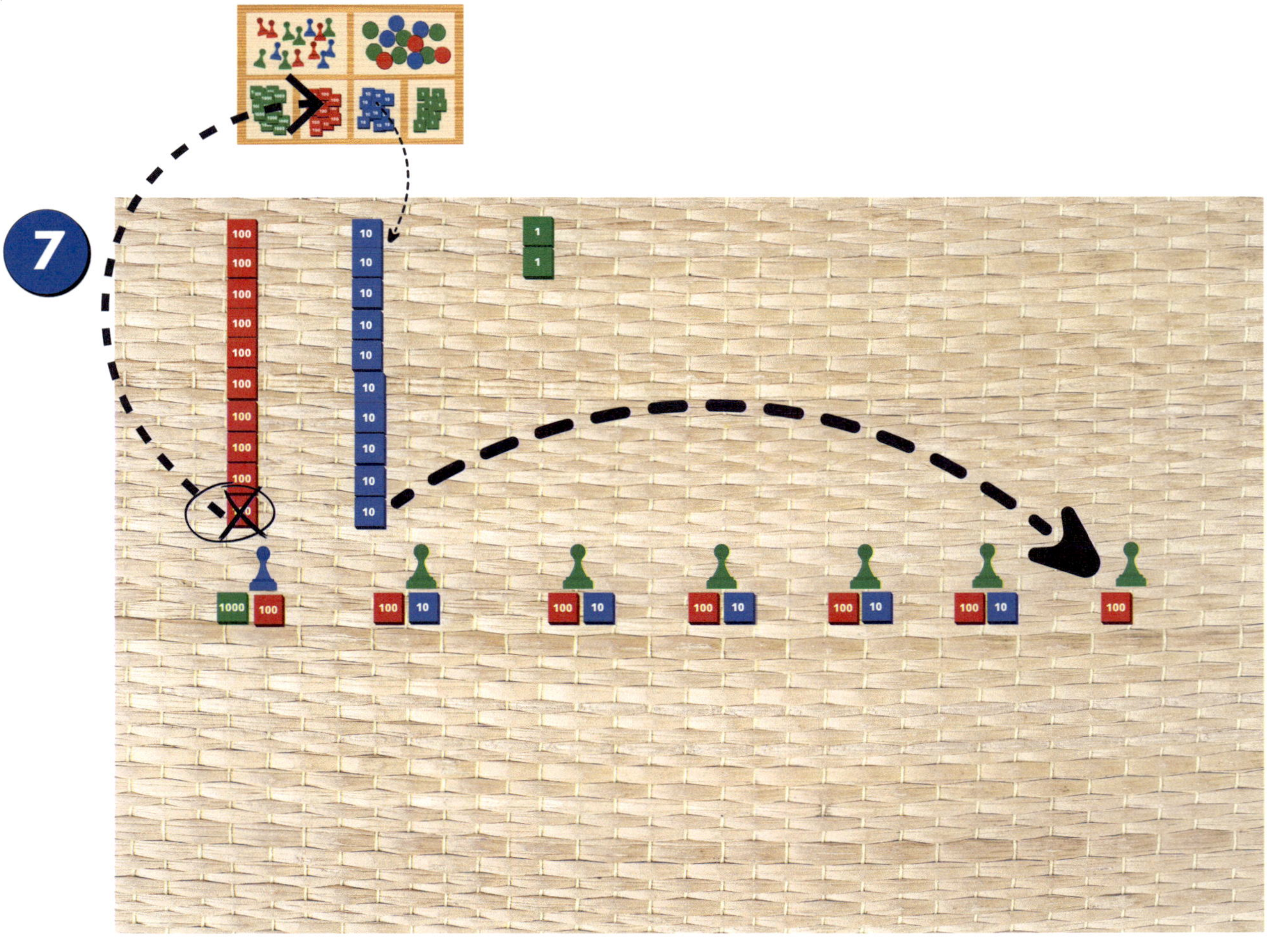

**8.** Continue distributing the stamps. Give one hundred stamp to the blue skittle followed by one ten stamp to each green skittle. Continue distributing, making exchanges as you go.

8

**9.** Make an exchange of one 100 for ten 10 stamps. Continue distributing.

9

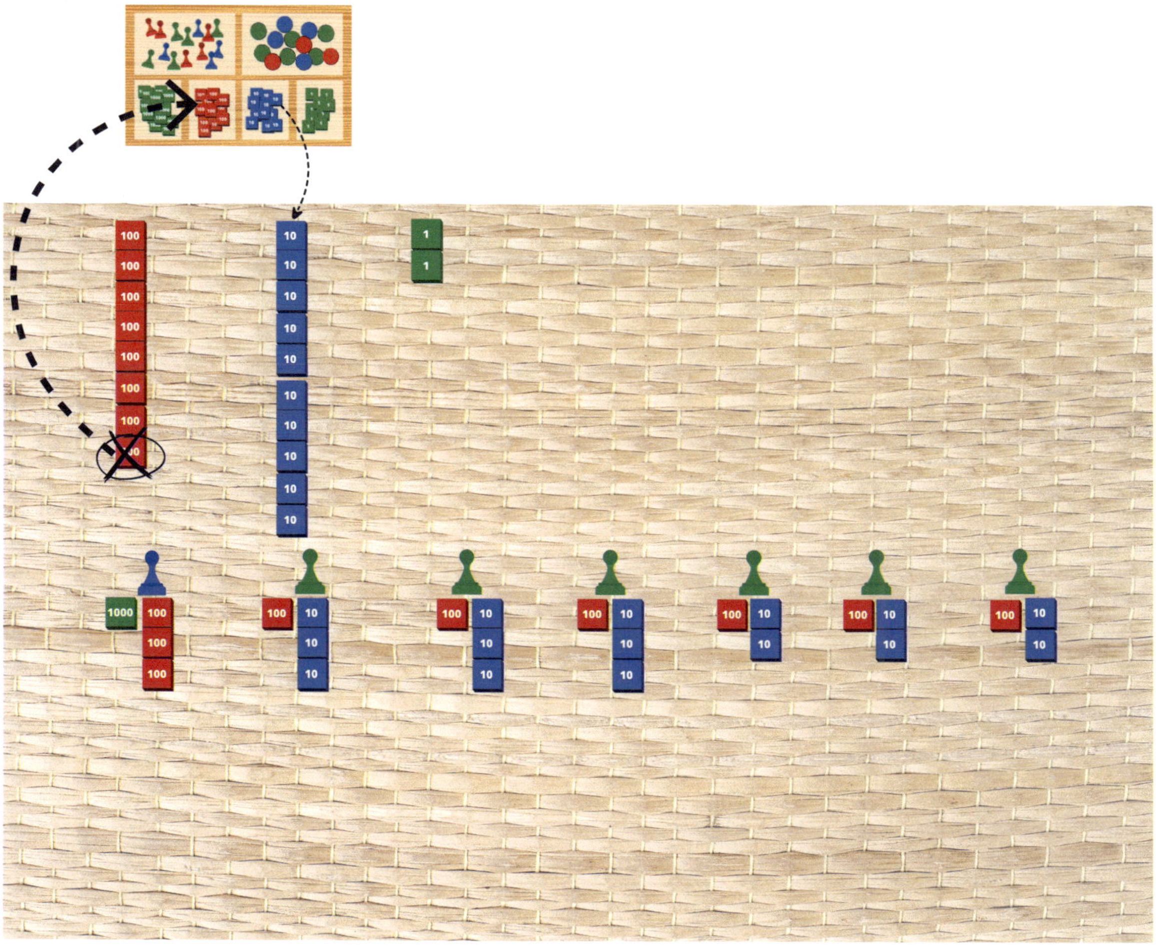

**10.** Make an exchange of one 100 stamp for ten 10 stamps. Continue distr buting.

**11.** Make another exchange. Continue distributing.

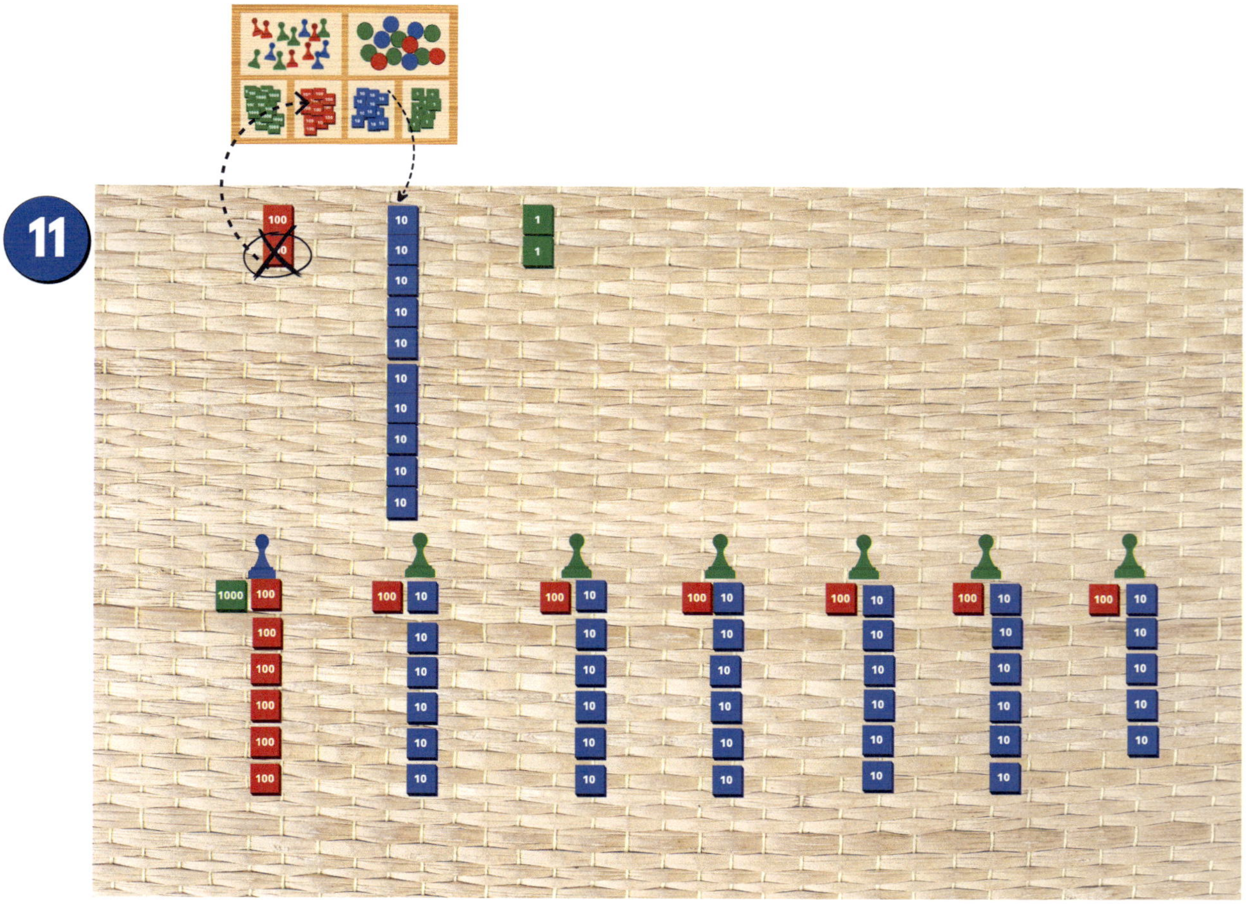

**12.** Now that the hundred stamps are distributed, continue with the ten stamps. Give one ten stamp to the blue skittle. Then distribute one unit stamp to each green skittle. Exchange as needed.

12

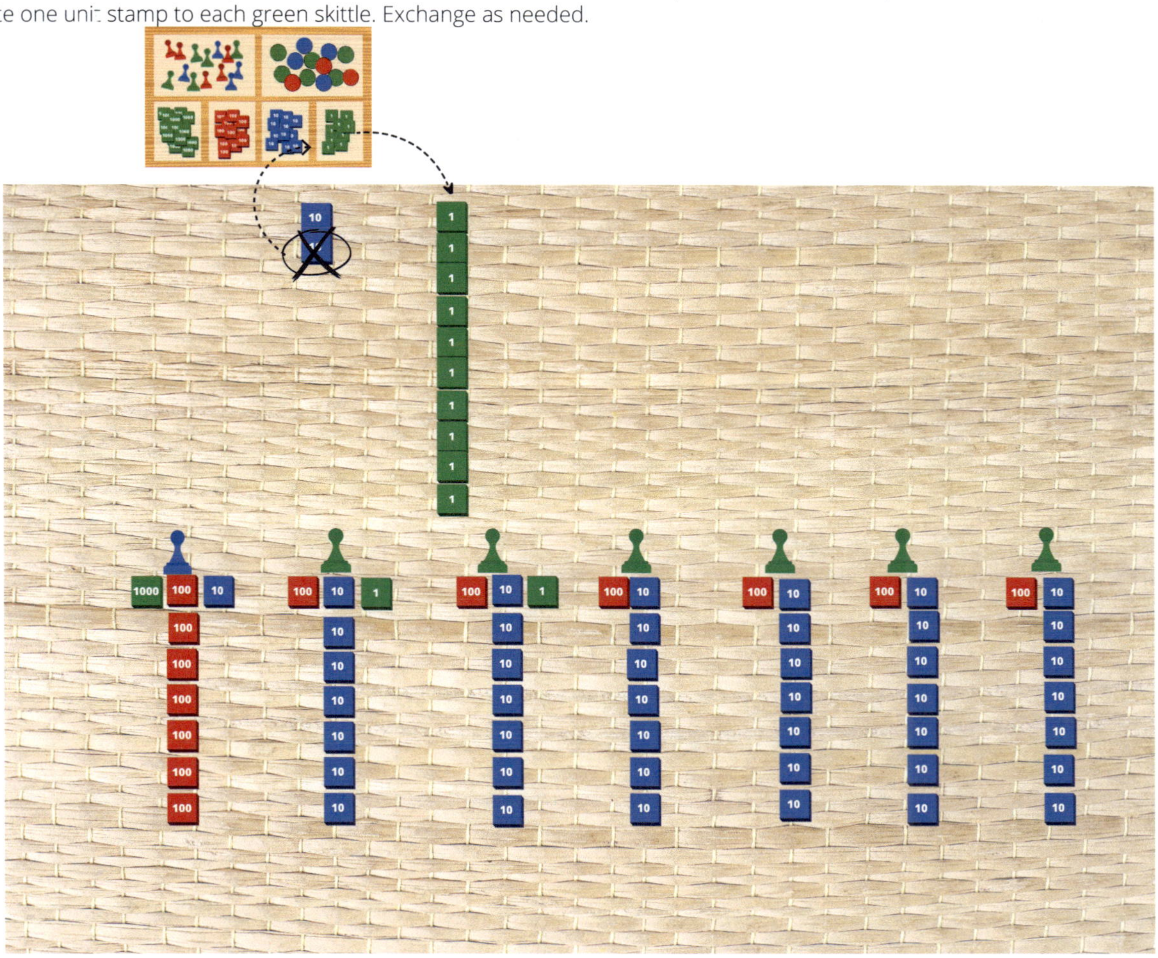

**13.** All of the stamps are now distributed! The answer is the the amount given to one stamp. [Answer: 172.]

13

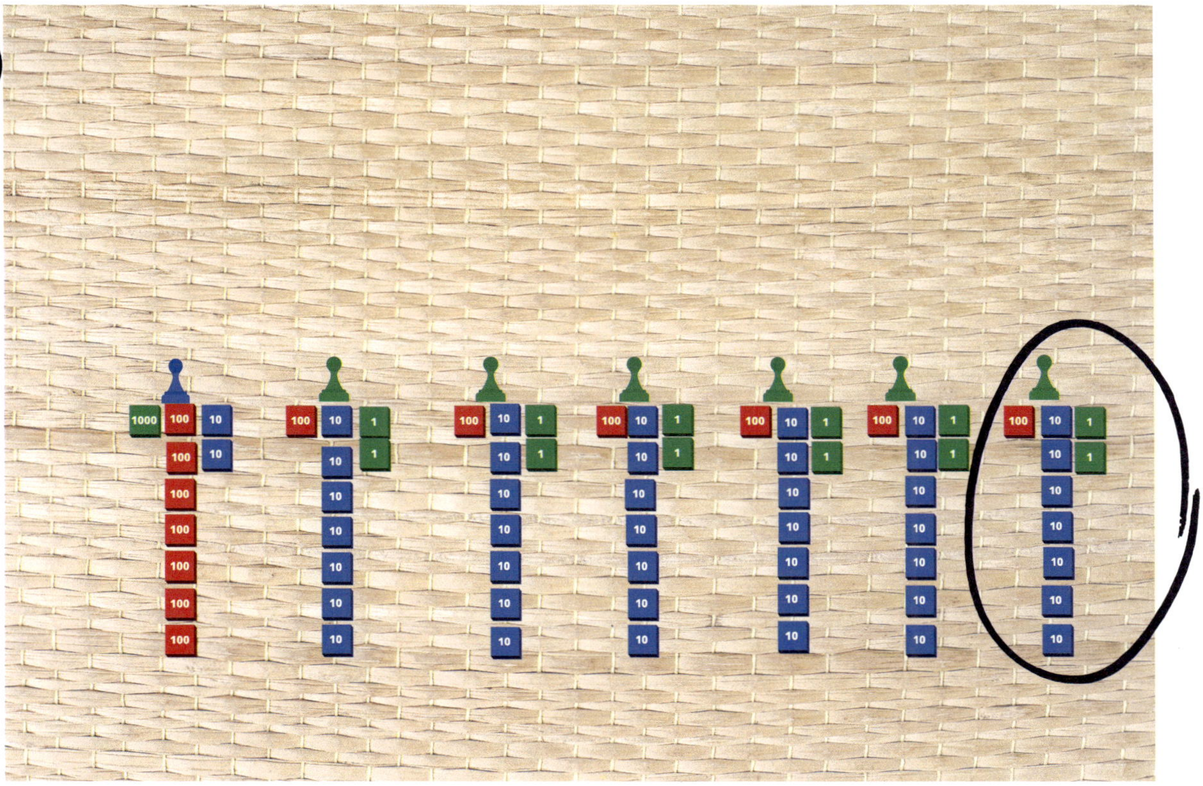

# 9.7 | Stamp Game: Division with 2-Digit Divisor - Advanced Problems

**Directions**:
Follow the directions leading up to this problem. You will need to make exchanges. Remember to use one blue skittle to represent each group of "ten people" and one green skittles to represent every "one person."

1. $2752 \div 16 =$ ____________________

2. $2628 \div 18 =$ ____________________

3. $3528 \div 14 =$ ____________________

4. $6372 \div 27 =$ ____________________

5. $9548 \div 44 =$ ____________________

6. $7956 \div 34 =$ ____________________

# 9.8 | Stamp Game: Division with Zero in Dividend

**Directions**:
1. Perform the division in the same manner as 2-digit division.
2. In order to get the necessary stamps under each skittle, you will need to make an exchange.
3. Write any remainders on the line. If there is no remainder, you can leave it blank or write "0."

1. 1 3 0 8 ÷ 1 2 = □□□□ R = ____

2. 2 8 5 0 ÷ 1 5 = □□□□ R = ____

3. 3 3 0 3 ÷ 2 2 = □□□□ R = ____

4. 4 0 6 5 ÷ 2 6 = □□□□ R = ____

# 9.9 | Stamp Game: Division with Zero in Divisor

**Background:**

1. Perform the division in the same manner as 2-digit division. You will use the red, blue, and green skittles or discs. If there is a *zero* in the divisor, then you need to use a *disc* for that category. **No stamps will be given to that "zero" category**.
2. In order to get the necessary stamps under each skittle, you may need to make an exchange.

1. 2 2 4 4 ÷ 1 0 2 = ☐ ☐ ☐ ☐

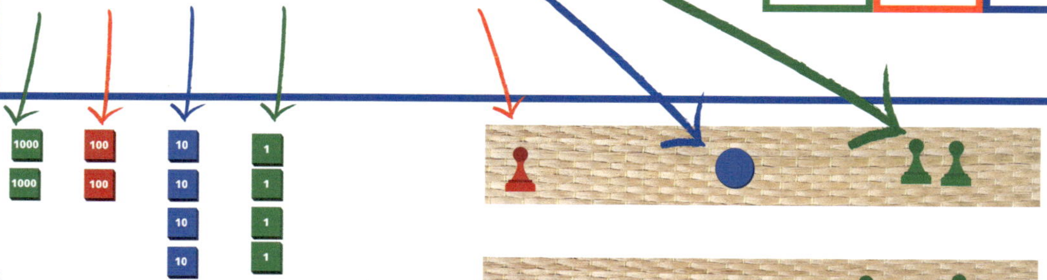


**Directions:**

1. First, lay out the stamps for the divisor and the skittles and disc for the dividend. Point out that the disc means you will *not* distribute stamps to that category.
2. Distribute one thousand stamp to the red skittle. Say, "Wait! There are no "people" in the hundreds category, so do we not give that category any stamps. Give the green skittles each a 10 stamp. Remember, if the red skittle gets 1000, the blue skittle *would* get 100, and the green skittle gets 10, as we go down in the decimal order of ten for each skittle category."
3. Now distribute the hundreds and units, again skipping the blue disc.
4. Write down your answer.

2. 8 8 4 4 ÷ 4 0 2 = ☐ ☐ ☐ ☐

3. 5 9 9 2 ÷ 1 0 7 = ☐ ☐ ☐ ☐

4. 5 4 2 7 ÷ 6 0 3 = ☐ ☐ ☐ ☐

# 9.10 | Stamp Game: Division with Zero in Unit of Divisor

**Directions**:

1. Perform the division in the same manner as 2-digit division. You will use the red, blue, and green skittles or discs. If there is a *zero* in the divisor, then you need to use a *disc* for that category. No stamps will be given to that category.
2. In order to get the necessary stamps under each skittle, you may need to make an exchange.
3. [**ALERT**!] When you have zero in the unit of the divisor, your final answer will need to be found by dividing the amount in the tens category by ten to find out how much each individual child receives.

1. 6 7 6 2 ÷ 1 3 0 = ☐☐☐☐ R = ____

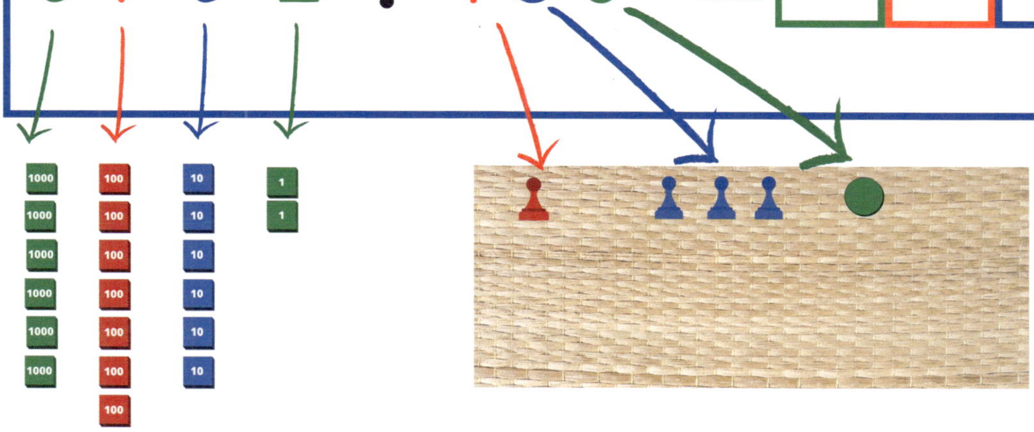

Remainder

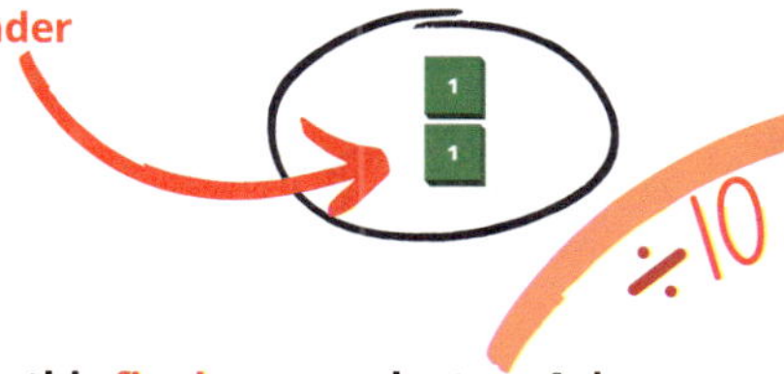


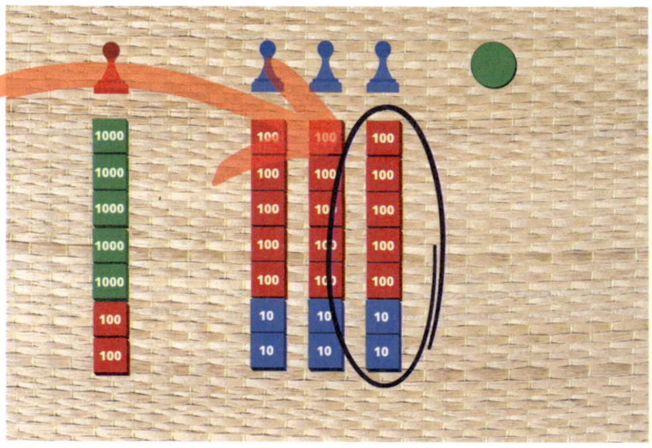

**Divide this final amount by ten. Ask, "What would each child get if this amount is shared by ten children?" You can perform another division problem with the stamps if this level of abstraction is too high for your student at this time.**

**Alternatively, you could find the next lower category for each stamp. Five hundred stamps would match with five ten stamps. Two ten stamps would match with two unit stamps. The final answer is 52. (Note the remainder of two that could not be distributed!)**

2. 9791 ÷ 110 = ☐☐☐☐ R = ____

3. 6155 ÷ 150 = ☐☐☐☐ R = ____

4. 3680 ÷ 160 = ☐☐☐☐ R = ____

# 9.11 | Stamp Game: Division Extension

**Directions**:

1. Color in the skittles you will need for this division problem. [Answer: one blue, four green.]
2. Perform the division. Since you are using the paper materials, put an X through a stamp once you distribute it. Circle *and* X the stamp if you exchange it. Use the cut-away Stamp Game to add more stamps. Just cut a strip that you need and glue it in the proper category with your dividend stamps.
3. To place a "Stamp" under a Skittle, you will color that square the proper color. Your answer is the amount under one green skittle at the end.

1 $448 \div 14 =$ ________________

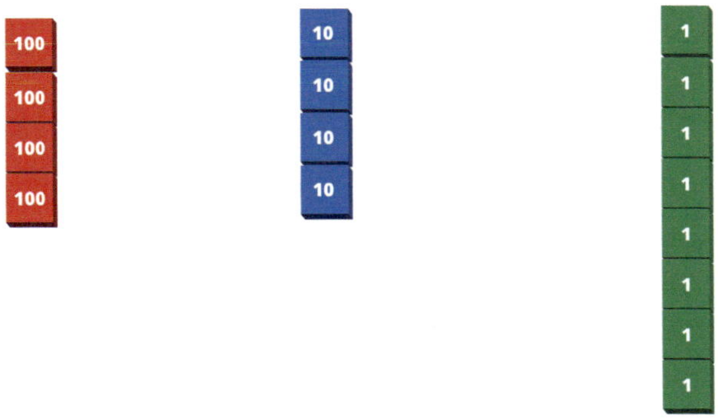


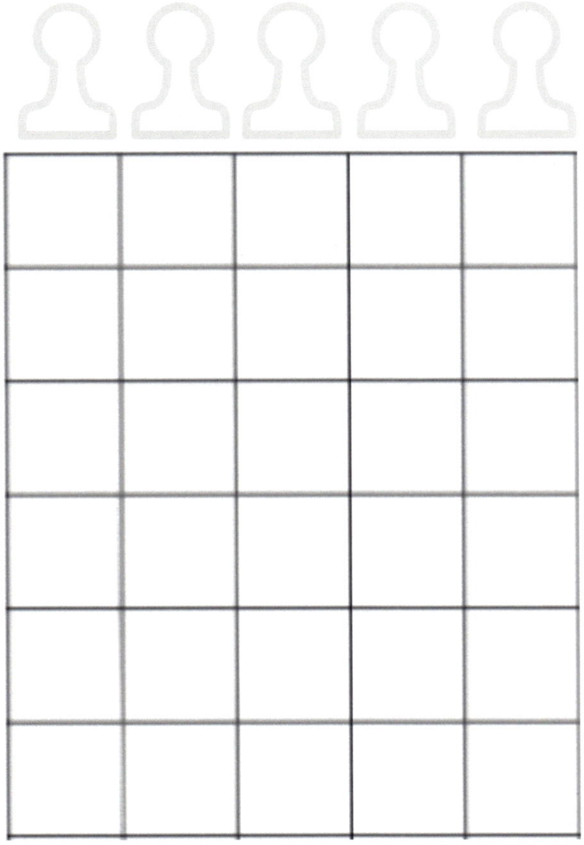

② $3696 \div 16 =$ ______________

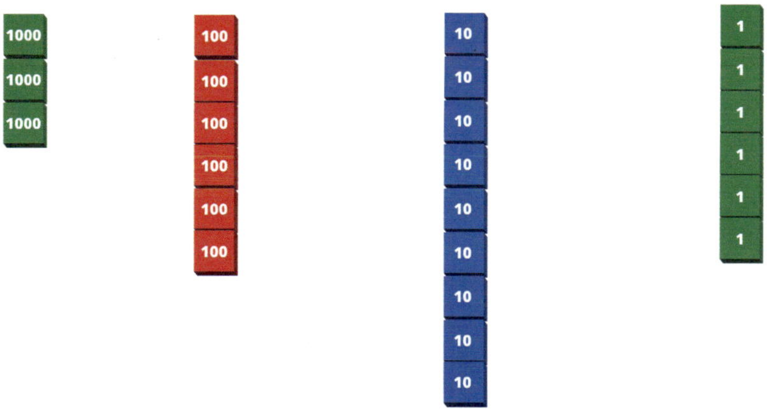
1000
1000
1000
100
100
100
100
100
100
10
10
10
10
10
10
10
10
10
1
1
1
1
1
1

**Directions:** Use the stamp game to solve. Check your work with your teacher and then color the picture.

A group of 8 kids got together to play marbles. They wanted to divide the marbles equally between them and leave out any extras. They had a pile of 150 marbles. How many marbles will each kid get? How many marbles will be left out of the game?

Each kid gets ____ marbles.

There are ____ marbles left out.

1 5 0 ÷ 8 = ____________

Jennie's mom made cookies for her dance class. She told Jennie to split them up between all of the kids equally and bring home the rest. There are 11 kids in Jennie's class, including Jennie. When Jennie opened the container, she counted 35 cookies. How many cookies does each child in the class get? How many cookies will Jennie bring home?

Each kid gets ____ cookies.

There are ____ cookies to bring home.

3 5 ÷ __ = ____________

Sarah, Casey, and Blake want to play a card game called "War." They need to distribute all of the cards in the deck between them. If any cards are leftover, those cards will be left out of the game. There are 52 cards in one deck of cards. How many cards will each player receive? How many cards will be left out of the game?

Each player gets ____ cards.

____ cards will be left out of the game.

__ __ ÷ __ = ____________

# 10.1 | Fractions Introduction

We will take some time now to introduce fractions before we talk about the clock. We will introduce the whole, half, and quarter.

**Directions:**

1. Cut out the circles on the bottom of this page. Cut apart the sections using the black lines as guides. Lay them spread out to the right of the workbook.
2. Point to the circle in the middle of this page. Say, **"This is a circle. This is a whole circle. This is one whole circle."**
3. Say, **"Let's find the whole circle."** Pick up the cut-out whole circle. Place the whole circle on top of the circle on the page to show they match. Set it as de
4. Say, **"Let's find a half circle."** Pick up one cut-out half circle. Place the half circle on top of the circle on the page. Leave it there.
5. Say, **"Let's find the other half circle."** Pick up the other cut-out half circle. Place it next to the other one on top of the circle on the page. Point to each half and say, **"Two half circles make one whole circle."** Take them off the page.
6. Pick up the quarter circle and name it. **"This is a quarter circle."** Superimpose the quarter circles on top of the full circle slowly and deliberately.
7. Count the quarters aloud. Say, **"There are four quarter-circles in one whole circle."**
8. Allow your child time to explore and make the circles using any parts he desires.

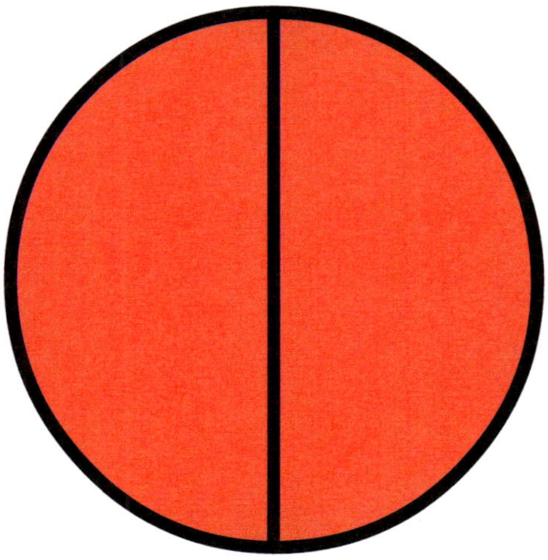

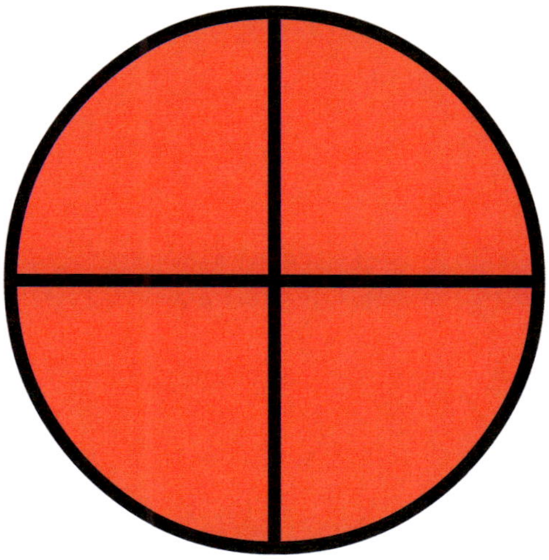

# 10.2 | Fractions Activity

**Background:**
In this activity, the student will be asked to color one half or one quarter of each shape a specific color. If your student is reading, he can read the directions himself. Non-readers will need you to read the directions aloud.
Encourage your student to use a pencil to "cut" the shape in half or quarters *before* coloring. Talk about ways to do this. If needed. make these shapes out of construction paper and *fold* the shapes to show symmetry.
***Troubleshooting***: Draw the lines for your student *before* asking him or her to color them if drawing symmetrical lines is too difficult at this time.

**Directions:**
1. Color one half of the heart red.
2. Color one quarter of the quatrefoil blue.
3. Color one half of the hexagon green.
4. Color one quarter of the ellipse pink.
5. Color one half of the triangle yellow.
6. Color one quarter of the rectangle black.
7. Color one half of the square purple.
8. Color one quarter of the circle orange.

# 10.3 | Skip Counting by 5's: Hundred Board

**Background**

Reading a clock will require the student to count by 5's and then add single units from there. For example, to show the time 6:17 on a clock, it is easiest to count 5-10-15 then 16, 17 for the minutes. So, today we will take the time to review counting by 5's. Then we will add a challenge by counting by 5's and continuing with units, using the hundred board as a visual aid.

**Directions:**

1. With a highlighter, circle the multiples of 5.
2. Read only the highlighted numbers when you finish.

(Note: If your child is not fluent in skip-counting by 5's, invite her to count 1-2-3-4-5 each time, and circle the number that she is on when she gets to "5.")

| | | | | | | | | | |
|---|---|---|---|---|---|---|---|---|---|
| 1 | 2 | 3 | 4 | 5 | 6 | 7 | 8 | 9 | 10 |
| 11 | 12 | 13 | 14 | 15 | 16 | 17 | 18 | 19 | 20 |
| 21 | 22 | 23 | 24 | 25 | 26 | 27 | 28 | 29 | 30 |
| 31 | 32 | 33 | 34 | 35 | 36 | 37 | 38 | 39 | 40 |
| 41 | 42 | 43 | 44 | 45 | 46 | 47 | 48 | 49 | 50 |
| 51 | 52 | 53 | 54 | 55 | 56 | 57 | 58 | 59 | 60 |
| 61 | 62 | 63 | 64 | 65 | 66 | 67 | 68 | 69 | 70 |
| 71 | 72 | 73 | 74 | 75 | 76 | 77 | 78 | 79 | 80 |
| 81 | 82 | 83 | 84 | 85 | 86 | 87 | 88 | 89 | 90 |
| 91 | 92 | 93 | 94 | 95 | 96 | 97 | 98 | 99 | 100 |

**Challenge**: Skip Count by 5's to get to the **number**. Use the marked hundred board above to visualize the numerals as you count.

1. **26**
2. **42**
3. **17**
4. **12**
5. **33**
6. **56**
7. **6**
8. **49**
9. **8**
10. **19**
11. **22**
12. **31**

# 10.4 | Skip Counting by 5's: Bead Bars

**Directions:**

1. Color the beads light blue as you count each bead. Write the number on the arrow as you get to each one.
2. Then, count by 5's using the numbers on the arrows.

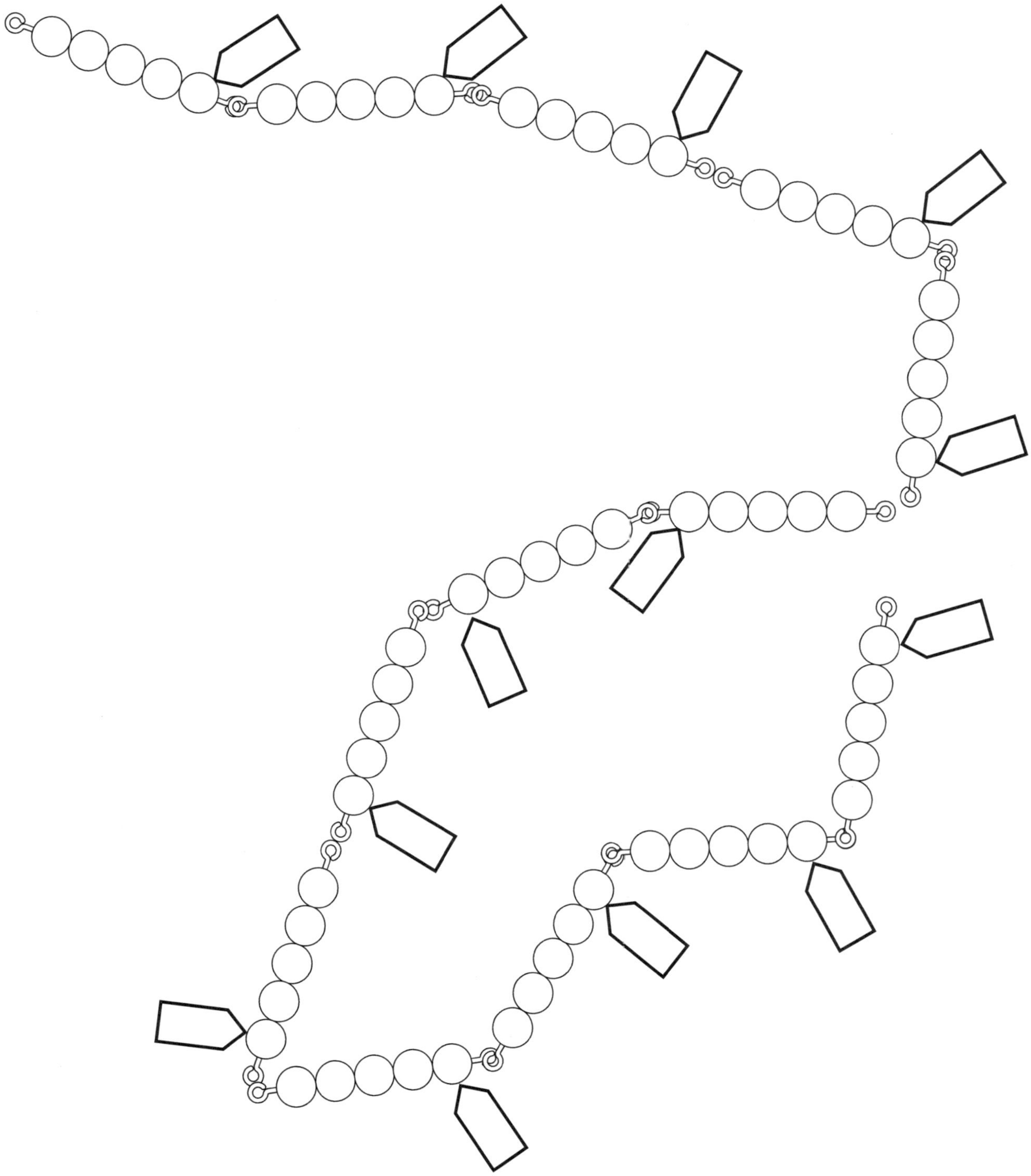

**Challenge**: Skip Count by 5's then count on by 1's to get to the **number**. Use the labeled beads above to visualize the numerals as you count. Label the bead with the corresponding letter.

A - **44** B - **18** C - **9** D - **57** E - **23**

# 10.5 | The Clock: Introduction

Now that you have reviewed *fractions* and *skip-counting by 5's,* you can move onto teaching your student about the analog clock.

## Part 1: A Real Analog Clock

Show your student a *real* analog clock.
Point out the following:

- The clock face is a circle.
- The numbers go around the clock in a "clockwise" direction. Point and count from 1-12.
- There are two *hands*. There is one short hand. That is the hour hand. There is one long hand. That is the minute hand.
- Note: If your clock has a second hand, show your student that it is the thinnest hand and moves every second. Have fun counting at the speed of the ticking. Explain we will not be using the second hand in our lessons.

## Part 2: Build a Clock

Using the printable clock materials provided in this lesson, you will build a demonstration clock with your student.

**Preparation (either with your student or ahead of time):**

1. Laminate the sheet with the clock and the clock components.
2. Cut out the clock components.
3. Apply adhesive putty, velcro dots, or rolled tape to the back of the red numerals.
4. Using a push pin and slice of cork (or a brass fastener), add the whole circle, the half circle, then the quarter circle, followed by the minute hand (blue), and lastly the hour hand (black) to the center of the clock. See if you can turn the parts. You may need to make the holes bigger if you are using a brass fastener.
5. Cut apart the bead bars. Add tape, adhesive putty, or pieces of velcro dots to the backs.
6. Disassemble before introducing the full lesson to your child.

**Directions:**

1. Using a real analog clock as a control (or the image on the next page), add the red numerals to the clock. (Note: Make sure you can remove these. See "Preparation" for ideas.)
2. Next, assemble the middle components. If you completed "Preparation," this will be demonstrating for your child the way you built the clock as before to make sure the parts are movable. Add the parts in the following order:
    1. Whole circle
    2. Half circle
    3. Quarter circle
    4. Minute hand
    5. Hour hand
    6. Bead bars
3. Next, point to each component, name it, and describe it, as follows:

(Note: Just give the quick description. There will be plenty of time to practice these concepts in the exercises and follow-up lessons!)

1. **Red Number Circles** - These are the numerals on the clock.
2. **Hour hand** - This is the hour hand. It points to the number of the hour.
3. **Minute hand** - This is the minute hand. It points to the number of minutes past the hour.
4. **Whole circle** - This is a whole circle. If the minute hand travels around the whole circle one time, then one hour has passed.
5. **Half circle -** This is a half circle. If the minute hand travels the half circle, then a half hour has passed.
6. **Quarter circle** - This is a quarter circle. If the minute hand travels the quarter circle, then a quarter hour has passed.
7. **Bead bars** - These are the bead bars of five. In between each red number circle on the clock, there are five minutes. When we put these all around the clock, we can find out how many minutes are in one hour! Ask, "How many minutes are in one hour?" (Allow child to count blue bead bars, preferably by skip counting by 5's.)

# Control Clock

This is the Control Clock. As you build your clock, use this page as a Control. (See directions on previous page.)

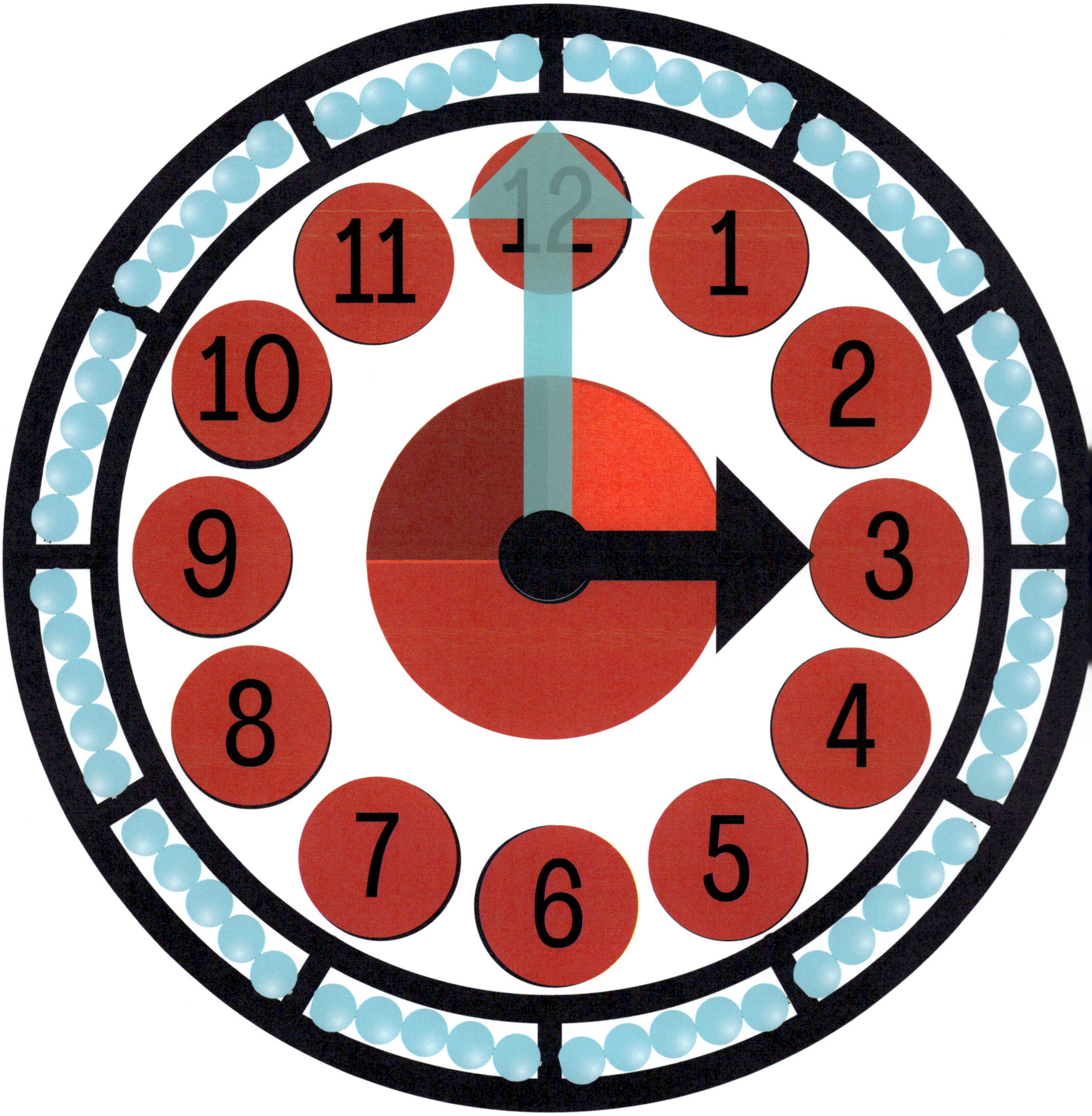

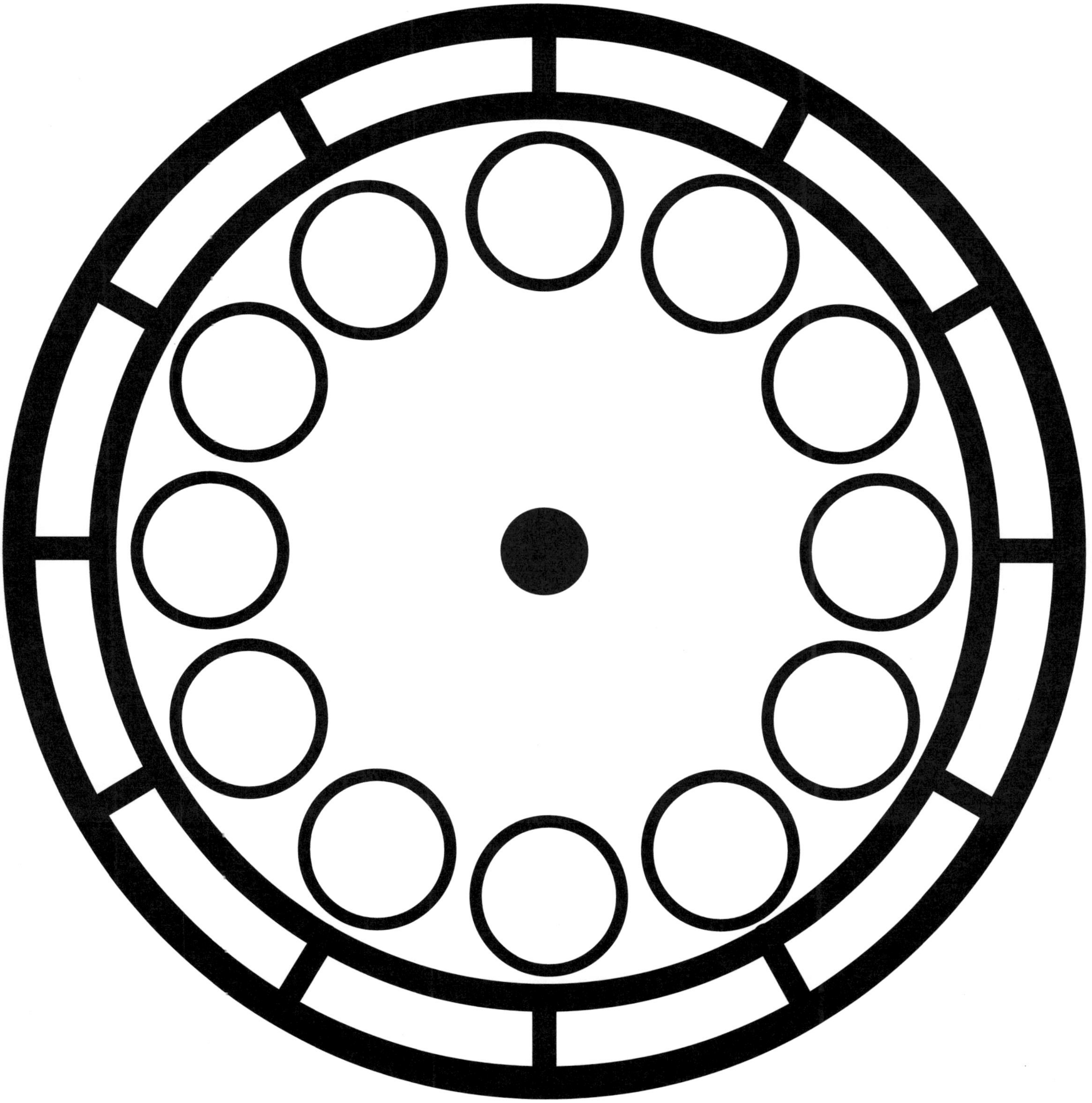

# Clock Components

See "The Clock: Introduction Lesson" for full instructions.

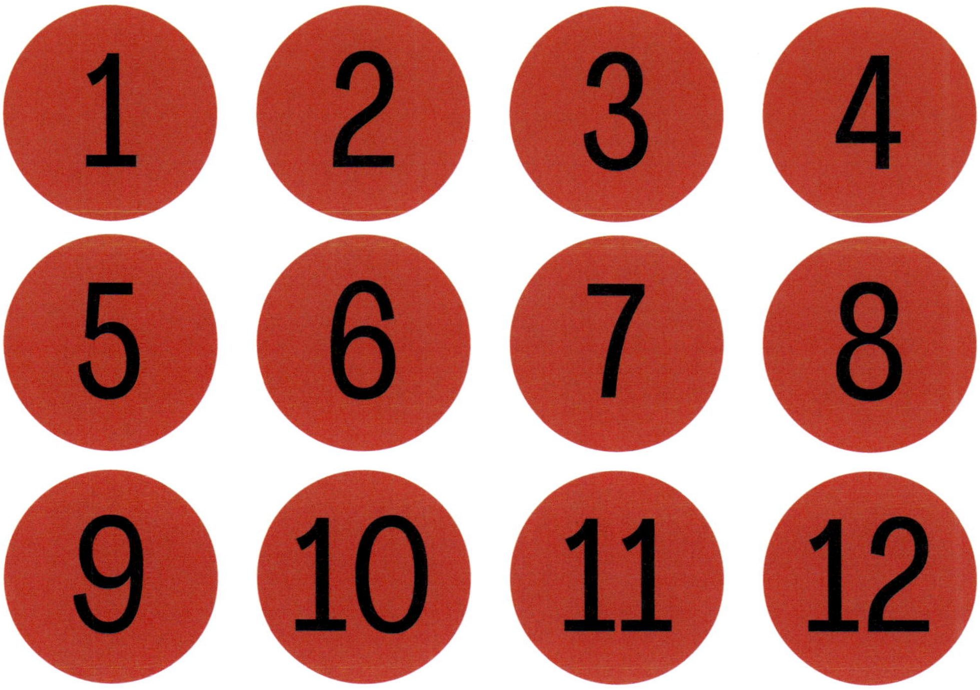

# 10.7 | The Clock: Telling Time to the Hour

**Directions:**

1. Say, "When we tell time to the hour, the long hand points to the "12" at the top of the clock, and the short hand points directly to the hour. We use the red whole circle on our movable clock. The minutes on the digital clock are "00." We also say "o'clock" when the clock is at exactly zero minutes past the hour."
2. Using the clock you built, practice showing your student how to make 1:00 to 12:00. Build the clock with the whole red circle only (no fraction circles.) Make the time, read the clock out loud, and write down the digital equivalent on a piece of small paper.
3. After you have made 1:00 through 12:00, invite your student to choose a slip of paper and make the time. Repeat as needed.
4. Complete this worksheet by filling in the missing digital times or clock hands using colored pencils.

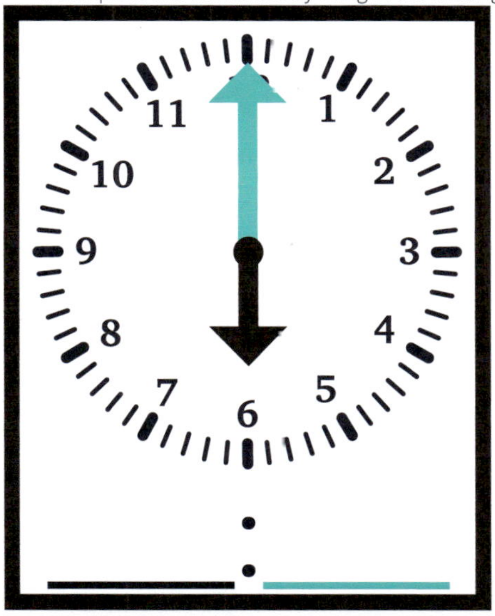

# 10.7 | The Clock: Telling Time to the Hour, Part 2

**Directions:** Match the clock face to the time by drawing a line to connect the matching times.

11:00

9:00

2:00

1:00

5:00

# 10.8 | The Clock: Telling Time to the Half Hour

**Directions:**

1. Say, "When we tell time to the half hour, the long hand points to the "6" at the bottom of the clock, and the short hand points halfway past the hour. We use the red half circle on our movable clock. The minutes on the digital clock are "30." We also say "thirty" or "half past" when the clock is at exactly halfway past the hour."
2. Using the clock you built, practice showing your student how to make 1:30 to 12:30. Build the clock with the half red circle only. Make the time, read the clock out loud, and write down the digital equivalent on a piece of small paper.
3. After you have made 1:30 through 12:30, invite your student to choose a slip of paper and make the time. Repeat as needed.
4. Complete this worksheet by filling in the missing digital times or clock hands using colored pencils.

____ : ____

____ : ____

____ : ____

____ : ____

____ : ____

____ : ____

4 : 30

10 : 30

6 : 30

# 10.9 | The Clock: Telling Time to the Quarter Hour

**Directions:**

1. Say, "When we tell time to the quarter hour, the long hand points to the "3" or "9", and the short hand points past the hour. We use the red quarter circle on our movable clock. When the minutes on the digital clock are "15," the long blue hand is on the "3." We say, for example, "six fifteen" or "quarter past six" when the clock is at exactly one quarter past the hour. When the minutes on the digital clock are "45," we say "six forty-five" or "a quarter til seven.""
2. Using the clock you built, practice showing your student how to make 1:00 to 12:45 in 15-minute increments. Build the clock with the quarter red circle, the half circle, and the whole circle. Make the time, counting by 5's and 15's, read the clock out loud, and write down the digital equivalent on a piece of small paper.
3. After you have made 1:00 through 12:45, invite your student to choose a slip of paper and make the time. Repeat as needed.
4. Complete this worksheet by filling in the missing digital times or clock hands using colored pencils.

____ : ____

____ : ____

____ : ____

____ : ____

____ : ____

____ : ____

7 : 15

8 : 45

6 : 15

# 10.10 | The Clock: Telling Time to the Minute

**Directions:**

1. Say, "When we tell time to the minute, the long hand points to any tick mark on the clock, and the short hand points to the hour or after the hour. We will use the whole circle, the half circle, and the quarter circle to help us determine the time. We will also use the five-bead bars."
2. Using the clock you built, practice showing your student how to make the times by counting by 5's and then counting on by 1's: 12:01, 1:05, 2:10, 3:11, 4:17, 5:20, 5:26, 6:31, 7:42, 8:46, 9:50, 10:55, 11:59. Read the clock time out loud, and write down the digital equivalent on a piece of small paper.
3. After you have made the given times, invite your student to choose a slip of paper and make the time. Repeat as needed.
4. Complete this worksheet by filling in the missing digital times or clock hands using colored pencils.

3 : 16

8 : 38

7 : 02

# 11.1 | The Calendar: 12 Months in a Year

The next few lessons will introduce the Calendar. You may already do a version of "Calendar" with your student, but if you do not, this will help get you into the routine. You will find Student Calendar Pages following this lesson. You will start to fill in the calendar at the current month. You can use the Student Calendar Pages to record the date each day you do math *or* each day you do school. You may even choose to put the calendar page in an accessible area (like on your fridge) and give your child the daily task of adding Today's Date to the calendar.

Also, if you want to make a reusable perpetual calendar, see instructions on Page **153.**

We will first discuss that there are 12 months in a year. You may want to elaborate by talking about seasons or mentioning when certain holidays fall, but I encourage you to keep the initial Calendar presentations simple.

**Months of the Year Song**
Tune: Ten Little Indians

*January*
*February*
*March*
*and April!*

*May*
*June*
*July*
*and August!*

*September*
*October*
*November*
*December*
*These are the months of the year!*

**BEFORE the Lesson PREP:**

1. Write the current year across the top of the first 2 Yearly Calendar pages.
2. Write the next year across the top of the second 2 Yearly Calendar pages (only if you will continue these activities into next year.)
3. Fill in the dates of the entire two years.

**Materials:**

- Yearly Calendars for this school year (filled out ahead of the lesson) (Page **144-147**)
- Blank Student Calendar Page (Page **159**)
- Sheet of month names (Page **157**)
- Sheet of month images (Page **155**)

**Months of the Year Song:**
https://youtu.be/5enDRrWyXaw

**Directions:**

1. Say to your student, "Today, we are in the year _______. There are 12 months in one year. Today, we are in the month of _____."
2. Count the months on the Yearly Calendar for *this year*.
3. Point out *this month* to your student.
4. Teach your child the **"Months of the Year"** song. See Lyrics and YouTube link above.
5. Invite your student to use a Student Calendar page, decoration, and name of the month to create this month's calendar. Add only the dates up to today's date. Every day you use the Calendar page, add only the dates up to the current date. Some days it will be one date and others you may have to "catch up" from a weekend or time off from using the Math Workbook.

*Yearly Calendar Pages (parent fills in dates as control)*

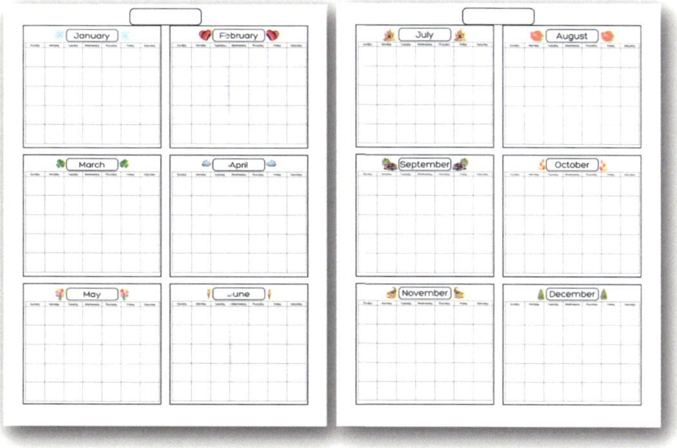


*Monthly Student Calendar Page:*

*There is extra space on each page to add your student's illustrations, notes, holidays, birthdays, or special events!*

## January

| Sunday | Monday | Tuesday | Wednesday | Thursday | Friday | Saturday |
|---|---|---|---|---|---|---|
| | | | | | | |
| | | | | | | |
| | | | | | | |
| | | | | | | |
| | | | | | | |

## February

| Sunday | Monday | Tuesday | Wednesday | Thursday | Friday | Saturday |
|---|---|---|---|---|---|---|
| | | | | | | |
| | | | | | | |
| | | | | | | |
| | | | | | | |
| | | | | | | |

## March

| Sunday | Monday | Tuesday | Wednesday | Thursday | Friday | Saturday |
|---|---|---|---|---|---|---|
| | | | | | | |
| | | | | | | |
| | | | | | | |
| | | | | | | |
| | | | | | | |

## April

| Sunday | Monday | Tuesday | Wednesday | Thursday | Friday | Saturday |
|---|---|---|---|---|---|---|
| | | | | | | |
| | | | | | | |
| | | | | | | |
| | | | | | | |
| | | | | | | |

## May

| Sunday | Monday | Tuesday | Wednesday | Thursday | Friday | Saturday |
|---|---|---|---|---|---|---|
| | | | | | | |
| | | | | | | |
| | | | | | | |
| | | | | | | |
| | | | | | | |

## June

| Sunday | Monday | Tuesday | Wednesday | Thursday | Friday | Saturday |
|---|---|---|---|---|---|---|
| | | | | | | |
| | | | | | | |
| | | | | | | |
| | | | | | | |
| | | | | | | |

# July

| Sunday | Monday | Tuesday | Wednesday | Thursday | Friday | Saturday |
|---|---|---|---|---|---|---|
| | | | | | | |
| | | | | | | |
| | | | | | | |
| | | | | | | |
| | | | | | | |

# August

| Sunday | Monday | Tuesday | Wednesday | Thursday | Friday | Saturday |
|---|---|---|---|---|---|---|
| | | | | | | |
| | | | | | | |
| | | | | | | |
| | | | | | | |
| | | | | | | |

# September

| Sunday | Monday | Tuesday | Wednesday | Thursday | Friday | Saturday |
|---|---|---|---|---|---|---|
| | | | | | | |
| | | | | | | |
| | | | | | | |
| | | | | | | |
| | | | | | | |

# October

| Sunday | Monday | Tuesday | Wednesday | Thursday | Friday | Saturday |
|---|---|---|---|---|---|---|
| | | | | | | |
| | | | | | | |
| | | | | | | |
| | | | | | | |
| | | | | | | |

# November

| Sunday | Monday | Tuesday | Wednesday | Thursday | Friday | Saturday |
|---|---|---|---|---|---|---|
| | | | | | | |
| | | | | | | |
| | | | | | | |
| | | | | | | |
| | | | | | | |

# December

| Sunday | Monday | Tuesday | Wednesday | Thursday | Friday | Saturday |
|---|---|---|---|---|---|---|
| | | | | | | |
| | | | | | | |
| | | | | | | |
| | | | | | | |
| | | | | | | |

## January

| Sunday | Monday | Tuesday | Wednesday | Thursday | Friday | Saturday |
|---|---|---|---|---|---|---|
| | | | | | | |
| | | | | | | |
| | | | | | | |
| | | | | | | |
| | | | | | | |

## February

| Sunday | Monday | Tuesday | Wednesday | Thursday | Friday | Saturday |
|---|---|---|---|---|---|---|
| | | | | | | |
| | | | | | | |
| | | | | | | |
| | | | | | | |
| | | | | | | |

## March

| Sunday | Monday | Tuesday | Wednesday | Thursday | Friday | Saturday |
|---|---|---|---|---|---|---|
| | | | | | | |
| | | | | | | |
| | | | | | | |
| | | | | | | |
| | | | | | | |

## April

| Sunday | Monday | Tuesday | Wednesday | Thursday | Friday | Saturday |
|---|---|---|---|---|---|---|
| | | | | | | |
| | | | | | | |
| | | | | | | |
| | | | | | | |
| | | | | | | |

## May

| Sunday | Monday | Tuesday | Wednesday | Thursday | Friday | Saturday |
|---|---|---|---|---|---|---|
| | | | | | | |
| | | | | | | |
| | | | | | | |
| | | | | | | |
| | | | | | | |

## June

| Sunday | Monday | Tuesday | Wednesday | Thursday | Friday | Saturday |
|---|---|---|---|---|---|---|
| | | | | | | |
| | | | | | | |
| | | | | | | |
| | | | | | | |
| | | | | | | |

# July

| Sunday | Monday | Tuesday | Wednesday | Thursday | Friday | Saturday |
|---|---|---|---|---|---|---|
| | | | | | | |
| | | | | | | |
| | | | | | | |
| | | | | | | |
| | | | | | | |

# August

| Sunday | Monday | Tuesday | Wednesday | Thursday | Friday | Saturday |
|---|---|---|---|---|---|---|
| | | | | | | |
| | | | | | | |
| | | | | | | |
| | | | | | | |
| | | | | | | |

# September

| Sunday | Monday | Tuesday | Wednesday | Thursday | Friday | Saturday |
|---|---|---|---|---|---|---|
| | | | | | | |
| | | | | | | |
| | | | | | | |
| | | | | | | |
| | | | | | | |

# October

| Sunday | Monday | Tuesday | Wednesday | Thursday | Friday | Saturday |
|---|---|---|---|---|---|---|
| | | | | | | |
| | | | | | | |
| | | | | | | |
| | | | | | | |
| | | | | | | |

# November

| Sunday | Monday | Tuesday | Wednesday | Thursday | Friday | Saturday |
|---|---|---|---|---|---|---|
| | | | | | | |
| | | | | | | |
| | | | | | | |
| | | | | | | |
| | | | | | | |

# December

| Sunday | Monday | Tuesday | Wednesday | Thursday | Friday | Saturday |
|---|---|---|---|---|---|---|
| | | | | | | |
| | | | | | | |
| | | | | | | |
| | | | | | | |
| | | | | | | |

# 11.2 | The Calendar: 7 Days in a Week

Today, you will teach your child a "Days of the Week" song. One fun way to do this is to sing the days of the week to the "Addams Family" tune. Here is a clip on YouTube: https://youtu.be/HtQcnZ2JWsY.

**Days of the Week Song**

*Days of the Week (clap clap)*
*Days of the Week (clap clap)*
*Days of the Week, Days of the Week, Days of the Week (clap clap)*
*There's Sunday and there's Monday*
*There's Tuesday and there's Wednesday*
*There's Thursday and there's Friday*
*And then there's Saturday*
*Days of the Week (clap clap)*
*Days of the Week (clap clap)*
*Days of the Week, Days of the Week, Days of the Week (clap clap)*

**Directions**:

1. Trace the names of the days of the week. (Optional: Choose a color for each day, and trace the words using those colors. You could choose the Montessori Bead Bar colors, like red for Sunday, the first day of the week, green for Monday, pink for Tuesday, etc.)
2. If your child is ready for further handwriting practice, invite him to copy the name of each day on the line.
3. Sing the "Days of the Week" song together.
4. Underline first letter of each day.
5. Look at your Student Calendar Page for this month. Find the letters that match each day of the week that your student underlined.

Sunday

Monday

Tuesday

Wednesday

Thursday

Friday

Saturday

# 11.3 | The Calendar: 28-31 Days in a Month

Today you will introduce the number of days in each month. Use the poem "Thirty Days Hath September" found written out on this page, or another version that you know. You can also introduce the trick where you count your knuckles and in-between your knuckles to keep track of the months that have 31 days and those that do not. (See image. Knuckle months are written in black and have 31 days. In-between knuckle months are written in red and do not have 31 days.)

**Directions:**

1. Recite the number of days in a month poem, "Thirty Days Hath September."
2. Optional: Show your student the "knuckles" trick. The black arrows and months have 31 days. The red arrows and months have 30 days or 28-29 days. (See image.)
3. Look at the Yearly Calendar, which serves as the child's control calendar.
4. Which month(s) have 28 or 29 days? (Fill in the number of days on the line for *this year* next to the months.)
5. Which months have 30 days? (Fill in the number 30 on the line for those months.)
6. Which months have 31 days? (Fill in the number 31 on the line for those months.)
7. Make sure you write the dates for this month on your calendar if you haven't already. Stop at today's date.
8. Look at this month's Calendar together. Ask: Today is ?, Yesterday was ?, Tomorrow is ? (Answers can include the name of the day and/or the date, depending on child's readiness.)

**Thirty Days Hath September**

Thirty days hath September,
April, June, and November,
All the rest have thirty-one,
Save February at twenty-eight,
But leap year, coming once in four,
February then has one day more.

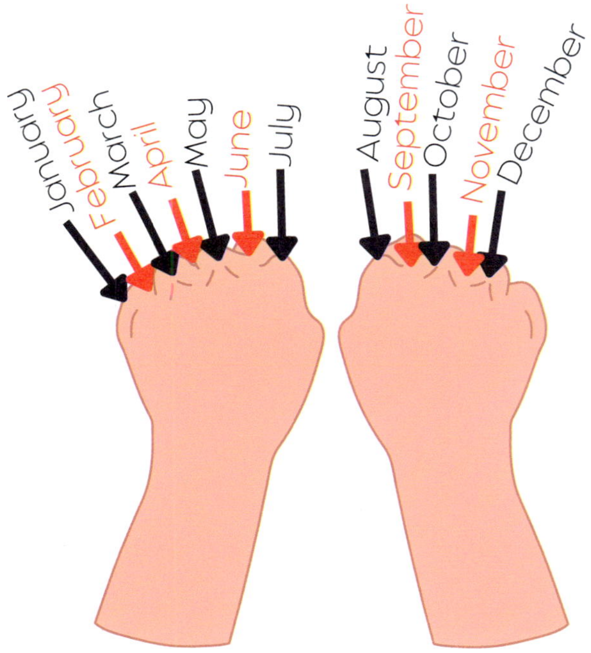


January: ______
February: ______
March: ______
April: ______
May: ______
June: ______
July: ______
August: ______
September: ______
October: ______
November: ______
December: ______

# 11.4 | The Calendar: Writing Today's Date

Today, you will show your student how to write today's date. Please modify this if you do not follow the same guidelines as shown. The way we write dates varies all over the world, which is a great lesson in itself!

**Directions:**

1. Say, "Today I am going to show you how to write today's date. To write the date, we need to know three things: the month, the day of the month, and the year."
2. Show your child this month's calendar. Find today. Ask, "What is the month?" (Help if needed.)
3. Ask, "What is the number of the day this month or *the date today?*" (Help if needed. Make sure to phrase the date as "The twenty-first," instead of "twenty-one.")
4. Ask, "What is the year?" (Help if needed. You can also add the year to the top of the monthly calendar as a reference.)
5. Say, "We now have enough information to write today's date. We start with the month, the date of the month, and then we write the year. I will show you." Write the date on the lines below.

____________________ ________, ______________

Month Date Year

6. Read the date again after writing it, pointing out the month, the date, and the year.
7. Ask your child to identify the month, date, and year. Then switch the order and repeat until solidified. (Note: if your child is having trouble identifying each term, put this lesson away and try again another day.)
8. Say, "That is one way to write the date. I am going to show you a shorter way to write the date. Instead of writing the name of the month, we are going to write the month number." Say the month out loud then count out loud from January while tallying the number on paper or on your fingers to determine the month number. Write the number on the line.
9. Say, "The date stays the same. Today is still the _______." Write today's date.
10. Say, "Now we get to shorten the way we write the date even more by abbreviating the year. We will write only the last two digits of the year." Write the last two digits of the year on the line.
11. Say, "When we do our math lessons from now on, we will date our work the short way."

________ / ________ / ________

Month Number Date 2-Digit Year

# 11.5 | The Clock and the Calendar: Ideas for Follow-Up

Below, there are some ways you can continue to practice the skills used in the Clock and Calendar lessons. Feel free to implement all or none of them, according to your child's ability or interest. All of these skills can be practiced orally, as well, if writing takes away from the core content of the lesson.

**Looping Review Exercises:**
Congratulations! You have completed the lessons in Units 10 and 11, and you are now ready to move on to completing the "hidden lesson" in each of the Looping Review Exercises! See the diagram below for each way these skills are practiced.

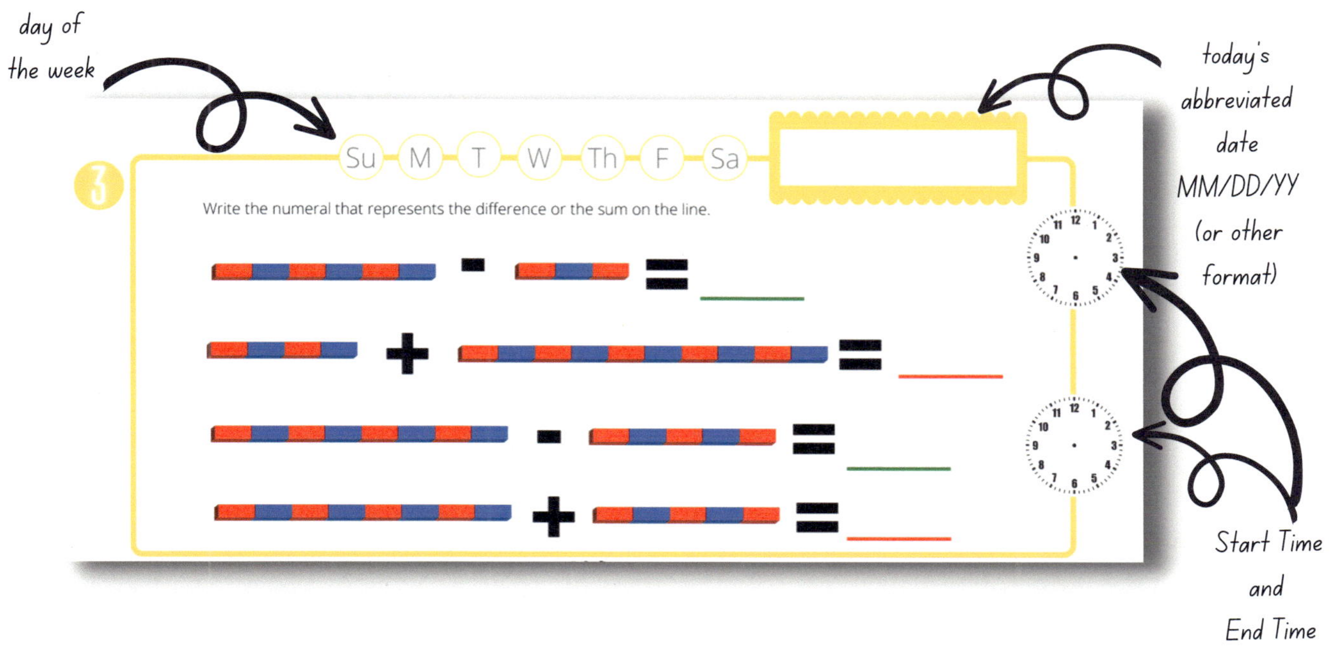


**Start of a Math Lesson:**
At the beginning of a regular math lesson or math work time, you can invite your student to write today's date, the time started (in an analog clock format like HH:MM), and the time finished.

**Daily Task:**
You can also have the Student Calendar page readily accessible for your student each day - in a work journal, on the fridge, in a clipboard hung on the wall, or the like.

**Wall Calendar:**
If writing the date is a hindrance or you have younger children in the mix, try a perpetual wall calendar with movable pieces for dates and month names! There are many options to purchase, and you can also make your own. You can laminate a single calendar page, the pictures for the months page, the names of the month, and the number squares provided on the next page. After laminating, attach small velcro dots to the back of each date square and within each section of the calendar.

# Number Squares for Perpetual Calendar

**Perpetual Calendar/Wall Calendar:**
If writing the date is a hindrance or you have younger children in the mix, try a perpetual wall calendar with movable pieces for dates and month names! There are many options to purchase, and you can also make your own.

You can laminate a single calendar page, the pictures for the months page, the names of the month, and the number squares provided on *this* page. After laminating, attach small velcro dots to the back of each date square and within each section of the calendar.

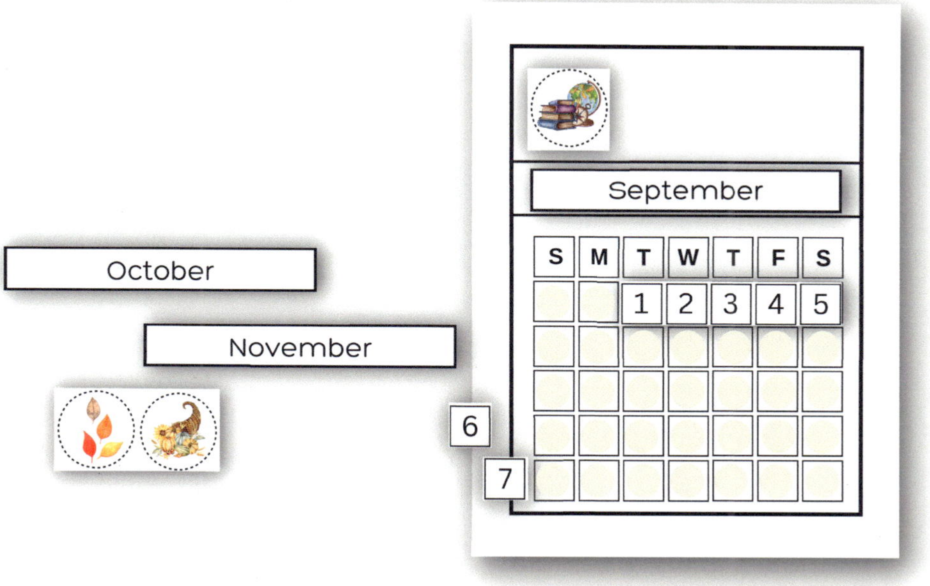


| | | | | | | |
|---|---|---|---|---|---|---|
| 1 | 2 | 3 | 4 | 5 | 6 | 7 |
| 8 | 9 | 10 | 11 | 12 | 13 | 14 |
| 15 | 16 | 17 | 18 | 19 | 20 | 21 |
| 22 | 23 | 24 | 25 | 26 | 27 | 28 |
| 29 | 30 | 31 | | | | |

| |
|---|
| January |
| February |
| March |
| April |
| May |
| June |
| July |
| August |
| September |
| October |
| November |
| December |

| S | M | T | W | T | F | S |
|---|---|---|---|---|---|---|
| | | | | | | |
| | | | | | | |
| | | | | | | |
| | | | | | | |
| | | | | | | |

| S | M | T | W | T | F | S |
|---|---|---|---|---|---|---|
| | | | | | | |
| | | | | | | |
| | | | | | | |
| | | | | | | |
| | | | | | | |

| S | M | T | W | T | F | S |
|---|---|---|---|---|---|---|
| | | | | | | |
| | | | | | | |
| | | | | | | |
| | | | | | | |
| | | | | | | |

| S | M | T | W | T | F | S |
|---|---|---|---|---|---|---|
| | | | | | | |
| | | | | | | |
| | | | | | | |
| | | | | | | |
| | | | | | | |

| S | M | T | W | T | F | S |
|---|---|---|---|---|---|---|
| | | | | | | |
| | | | | | | |
| | | | | | | |
| | | | | | | |
| | | | | | | |

| S | M | T | W | T | F | S |
|---|---|---|---|---|---|---|
| | | | | | | |
| | | | | | | |
| | | | | | | |
| | | | | | | |
| | | | | | | |

| S | M | T | W | T | F | S |
|---|---|---|---|---|---|---|
| | | | | | | |
| | | | | | | |
| | | | | | | |
| | | | | | | |
| | | | | | | |

| S | M | T | W | T | F | S |
|---|---|---|---|---|---|---|
| | | | | | | |
| | | | | | | |
| | | | | | | |
| | | | | | | |
| | | | | | | |

| S | M | T | W | T | F | S |
|---|---|---|---|---|---|---|
| | | | | | | |
| | | | | | | |
| | | | | | | |
| | | | | | | |
| | | | | | | |

| S | M | T | W | T | F | S |
|---|---|---|---|---|---|---|
| | | | | | | |
| | | | | | | |
| | | | | | | |
| | | | | | | |
| | | | | | | |

| S | M | T | W | T | F | S |
|---|---|---|---|---|---|---|
| | | | | | | |
| | | | | | | |
| | | | | | | |
| | | | | | | |
| | | | | | | |

| S | M | T | W | T | F | S |
|---|---|---|---|---|---|---|
| | | | | | | |
| | | | | | | |
| | | | | | | |
| | | | | | | |
| | | | | | | |

# Looping Review Exercises

As children progress through the Montessori math sequence, they will continually review concepts previously mastered as they work with new material. To keep those math muscles even stronger, you may consider inviting your child to go back and review prior lessons. Here you will find 60 days worth of "review" lessons that incorporate concepts from the Montessori Math Workbooks: Primary Book 1, Primary Book 2, and some new lessons from this book.

These exercises are set up to be completed one at a time. To hone additional math skills, invite your child to shade in the day of the week bubble, write the date in the blank "tab" box, and fill in the time your child starts the activity and then the time your child ends the activity. Depending on the age, ability, and interest of your child, you may want to start off filling in the times for your child. When he/she shows interest, invite your child to take over the job! This Workbook is unique in that it has specific lessons on writing the time and the date in Units 10 and 11. You may want to complete Units 10 and 11 at the same time as you are working through the Math Workbook 3. Once you have completed Units 10 and 11, you can move on to the Looping Review Exercises where your student will be more confident in writing the date and time. *(See **page iii** in the front of the Workbook for more detailed schedule suggestions.)*

The Looping Review Exercises do not require cutting and pasting or outside materials. Your child can complete these exercises with just a pencil and crayons (or colored pencils).

Your child will review previous math skills and newly acquired math skills as he/she loops through the skills twice.

**Some Skills that will be reviewed in these exercises are:**

- writing numbers and number words

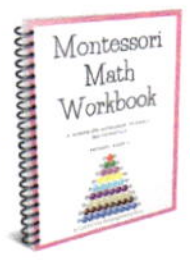


- number rods and operations with number rods
- even and odd
- greater than and less than
- the short bead stair and operations with the short bead stair
- teen number recognition and formation
- counting by tens
- counting 1-100 using the hundred chain and the hundred board
- squaring chains and skip counting

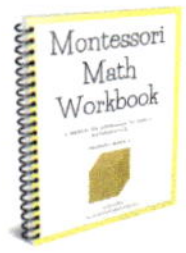


- the decimal system using the Golden Bead Material
- formation of numerals using the Golden Bead Material and decimal cards
- addition, subtraction, multiplication, and division using the Golden Bead Material
- addition strip board problems

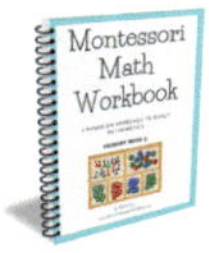


- addition working chart problems
- subtraction strip board problems
- bead houses

# Looping Review Exercises

1

Su M T W Th F Sa

1. Trace the numerals in your best handwriting.
2. Trace the words for 1, 2, and 3 on the line.
3. Draw a line from the words to the matching numerals.

0 1 2 3 4 5 6 7 8 9

one two three

2

Su M T W Th F Sa

1. Color the number rod.
2. Draw the rod that is two more than the given rod *above* the given rod.
3. Draw the rod that is one less than the given rod *below* the given rod.
4. Put a star to the *right* of the *greatest* rod.
5. Put a happy face to the *right* of the *least* rod.

3

Su M T W Th F Sa

Write the numeral that represents the difference or the sum on the line.

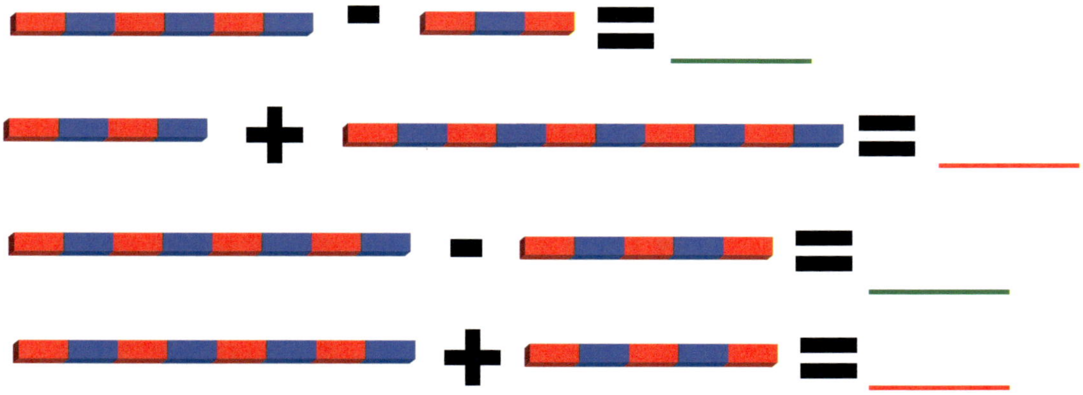

4

Su M T W Th F Sa

**Background:**
The only place value that needs to be examined to determine if a number is even or odd is the units place.
If a number has a 0, 2, 4, 6, or 8 in the units place, the number is *even*.
If a number has a 1, 3, 5, 7, or 9 in the units place, the number is *odd*.
Remember, if the units can be *paired*, the number is even.
If the units have an "odd one out" by itself, the number is odd.

**Directions:**
Draw the dots in pairs to represent only the *units* in the following numbers. Then circle the word "even" or "odd."

| 15 | 732 | 3216 |
|---|---|---|
| even odd | even odd | even odd |

5

Su M T W Th F Sa

**Make Ten Snake Game** (without cutting and pasting)
Count to ten. Draw a pencil mark after the tenth bead. Cross out the ten beads you counted and draw a golden bead ten-bar to start your new snake. Keep counting from your pencil mark until you get to 10. Draw a pencil mark after the tenth bead. Cross out the ten beads you counted and draw another golden bead ten-bar on your new snake. If there are less than ten beads at the end, draw the proper black and white bead bar. Bonus: Write out the addition equation!

# Looping Review Exercises

## 6

Su M T W Th F Sa

Show the teen numbers 11-19 by drawing the colored bead bars and writing the numerals in order from least to greatest. The first one is done for you.

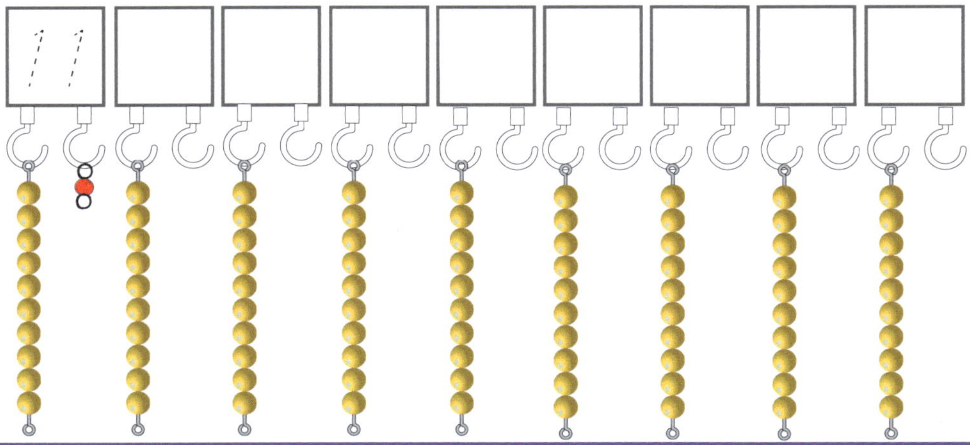

## 7

Su M T W Th F Sa

Draw a line from the numeral tile to the space it belongs on the teen board for numbers 11-15.

Then, write in the numerals needed to make 11-15 on the other board.

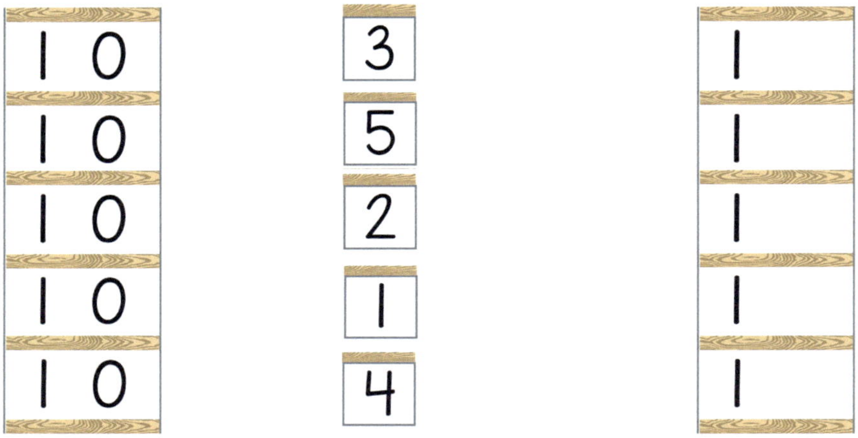

## 8

Su M T W Th F Sa

Write the teen numeral made by combining the rod of ten with another number rod.

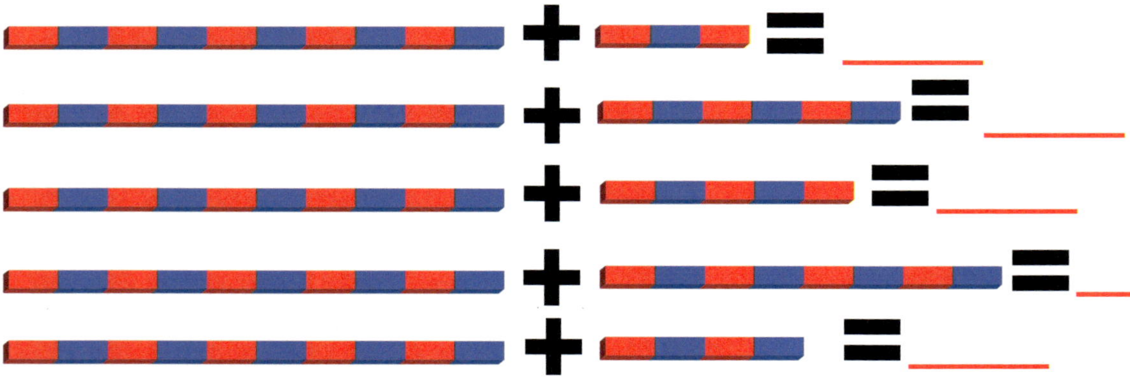

9

Su M T W Th F Sa

Color the beads that represent each numeral.

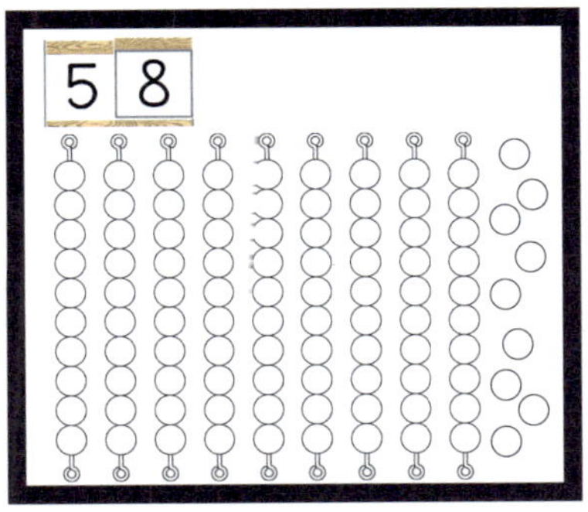


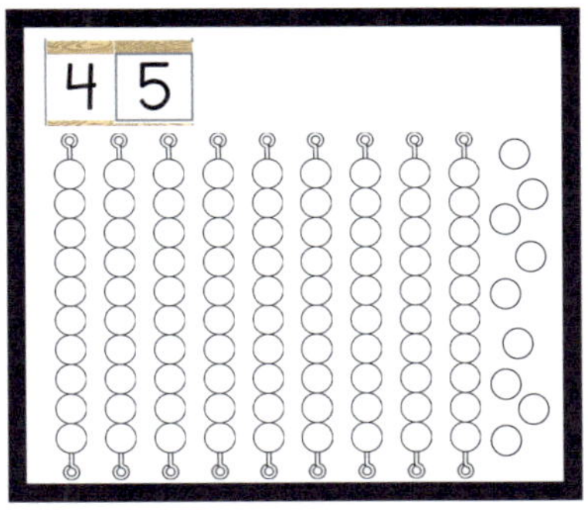


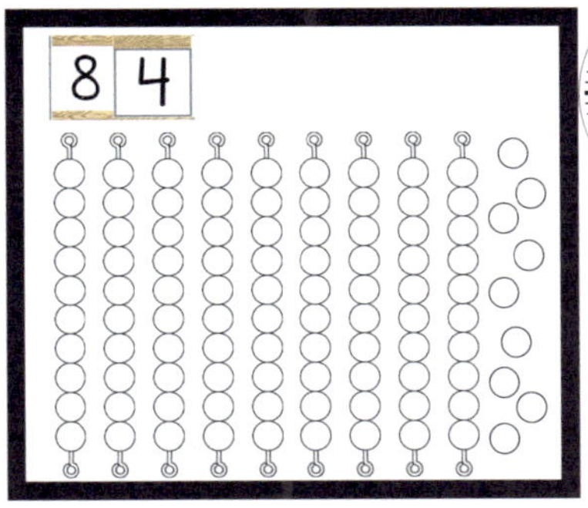

Write the numeral that represents each group of beads.

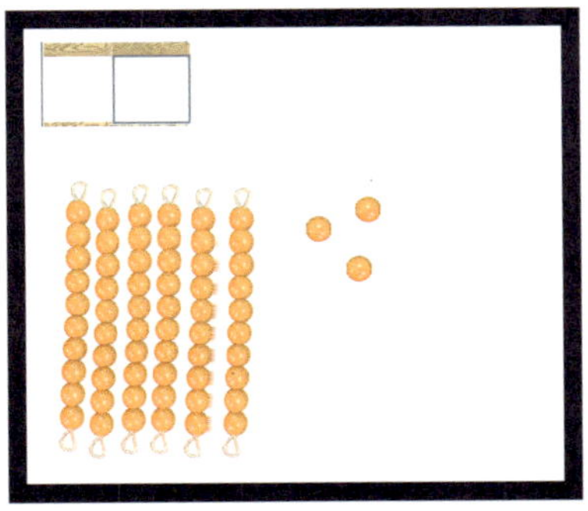

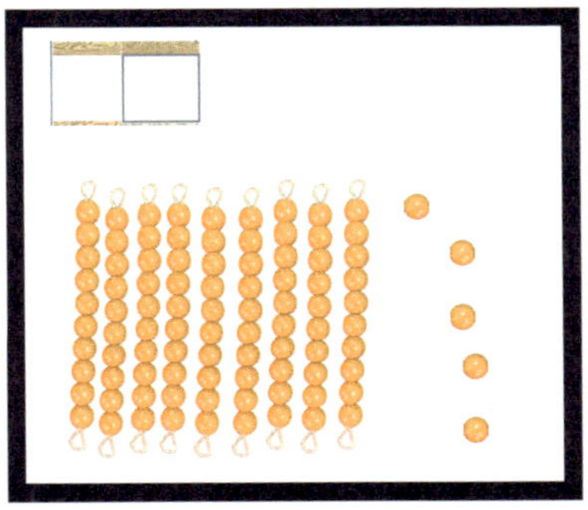

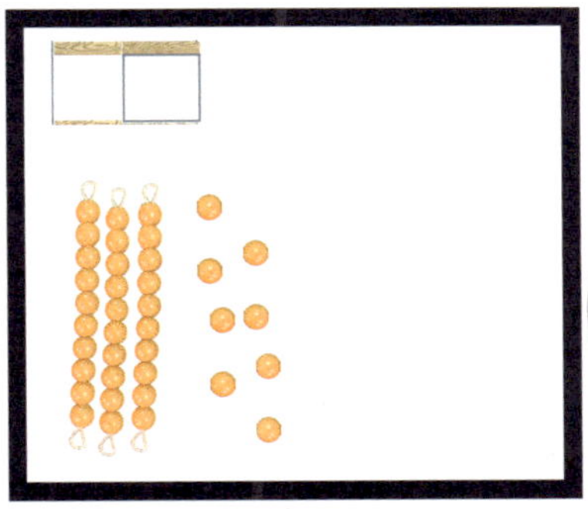

10

Su M T W Th F Sa

Color the beads along the 100 Chain according to the key below the Chain. The first is colored for you.

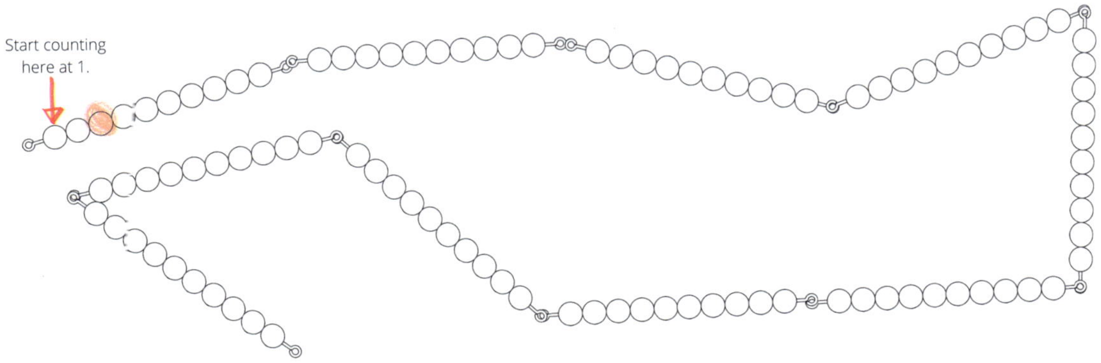


**3 = red** **27 = blue** **45 = green** **89 = yellow** **97 = pink**

## 11

Su M T W Th F Sa

Draw a line from the number to the spot on the hundred board where it belongs.

49
7
21
28
42
70
56
97
14
63
35

| | | | | | | | | | |
|---|---|---|---|---|---|---|---|---|---|
| 1 | 2 | 3 | 4 | 5 | 6 | | 8 | 9 | 10 |
| 11 | 12 | 13 | | 15 | 16 | 17 | 18 | 19 | 20 |
| | 22 | 23 | 24 | 25 | 26 | 27 | | 29 | 30 |
| 31 | 32 | 33 | 34 | | 36 | 37 | 38 | 39 | 40 |
| 41 | | 43 | 44 | 45 | 46 | 47 | 48 | | 50 |
| 51 | 52 | 53 | 54 | 55 | | 57 | 58 | 59 | 60 |
| 61 | 62 | | 64 | 65 | 66 | 67 | 68 | 69 | |
| 71 | 72 | 73 | 74 | 75 | 76 | 77 | 78 | 79 | 80 |
| 81 | 82 | 83 | 84 | 85 | 86 | 87 | 88 | 89 | 90 |
| 91 | 92 | 93 | 94 | 95 | 96 | | 98 | 99 | 100 |

## 12

Su M T W Th F Sa

Fill in the numbers of the beads on the squaring chain. Count the beads in the square.

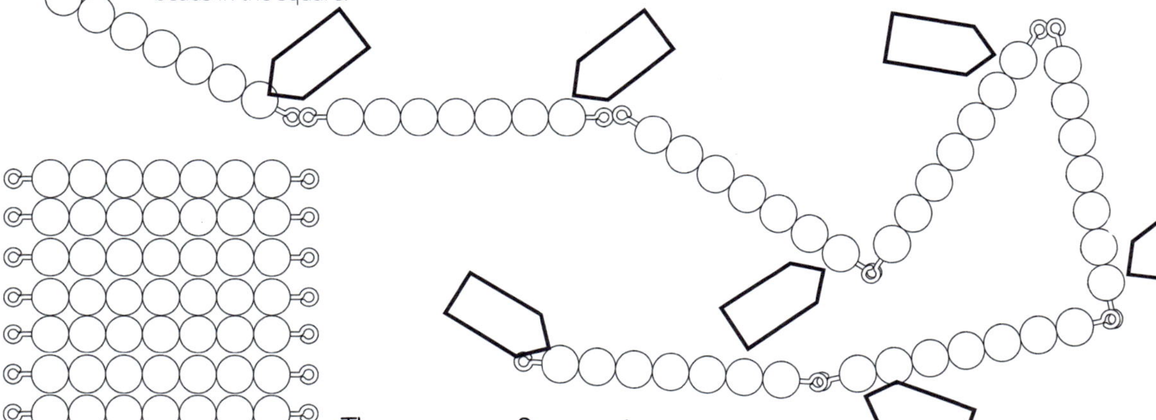

The square of seven is _____ .

## 13

Su M T W Th F Sa

Highlight or circle the multiples of three. (Skip count by 3's.)

| | | | | | | | | | |
|---|---|---|---|---|---|---|---|---|---|
| 1 | 2 | 3 | 4 | 5 | 6 | 7 | 8 | 9 | 10 |
| 11 | 12 | 13 | 14 | 15 | 16 | 17 | 18 | 19 | 20 |
| 21 | 22 | 23 | 24 | 25 | 26 | 27 | 28 | 29 | 30 |
| 31 | 32 | 33 | 34 | 35 | 36 | 37 | 38 | 39 | 40 |
| 41 | 42 | 43 | 44 | 45 | 46 | 47 | 48 | 49 | 50 |
| 51 | 52 | 53 | 54 | 55 | 56 | 57 | 58 | 59 | 60 |
| 61 | 62 | 63 | 64 | 65 | 66 | 67 | 68 | 69 | 70 |
| 71 | 72 | 73 | 74 | 75 | 76 | 77 | 78 | 79 | 80 |
| 81 | 82 | 83 | 84 | 85 | 86 | 87 | 88 | 89 | 90 |
| 91 | 92 | 93 | 94 | 95 | 96 | 97 | 98 | 99 | 100 |

14

Su M T W Th F Sa

How many  are in  ?______

How many  are in  ?______

How many 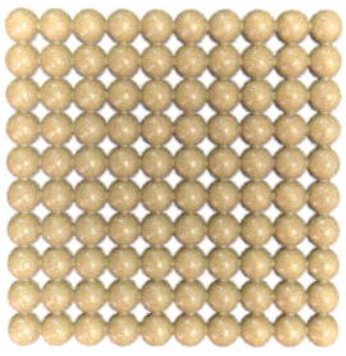 are in 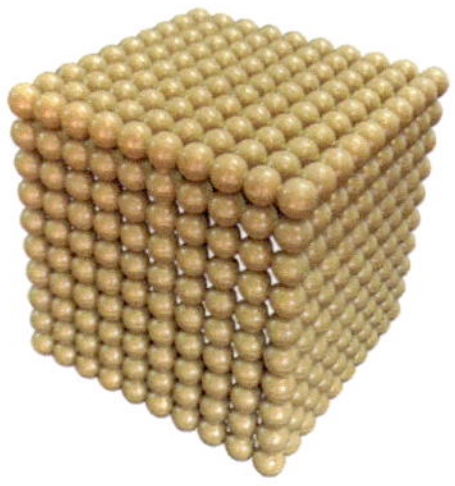 ?______

How many beads are in the chain? (Count by 10's.)__________

15

Su M T W Th F Sa

Draw a line from the number to the dot below the part of the mat where it belongs. Then write the proper numeral in each white box. (OR write the words representing each numeral.)

100 1000 10 1

# Looping Review Exercises

## 16

Su M T W Th F Sa

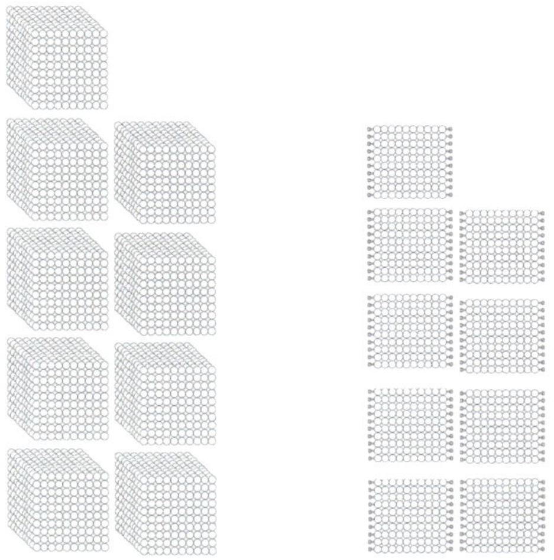

Color the corresponding amount of beads.

6 4 8 3

## 17

Su M T W Th F Sa

Circle or shade in the corresponding decimal cards for this amount. Then write the number.

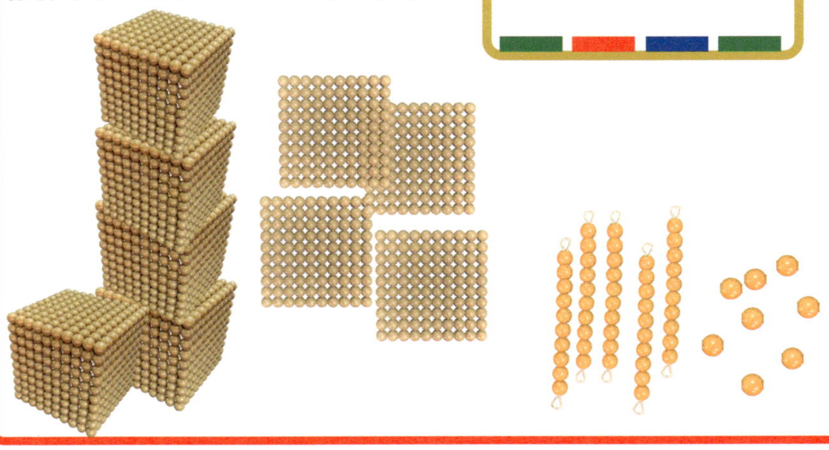

| | | | |
|---|---|---|---|
| 1000 | 100 | 10 | 1 |
| 2000 | 200 | 20 | 2 |
| 3000 | 300 | 30 | 3 |
| 4000 | 400 | 40 | 4 |
| 5000 | 500 | 50 | 5 |
| 6000 | 600 | 60 | 6 |
| 7000 | 700 | 70 | 7 |
| 8000 | 800 | 80 | 8 |
| 9000 | 900 | 90 | 9 |

## 18

Su M T W Th F Sa

Using the simple drawing method for the Golden Beads, draw out the problem on the addition mat. When you move the beads "down" under the red addition line, redraw them in the lower section and cross them out from the upper sections.Write the numeral answer in the bottom left box.

| | | | | | |
|---|---|---|---|---|---|
| | 1125 | 1000 | 100 | 20 | 5 |
| + | 3343 | 3000 | 300 | 40 | 3 |
| | | | | | |

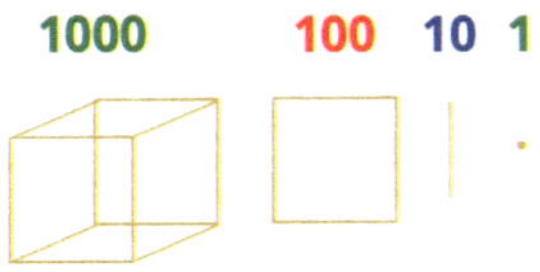

19

Su M T W Th F Sa

1. Write the numerals in the left column for the minuend and the subtrahend.
2. Cross out the beads from the top row (the minuend) that have been "moved down" to the middle row (the subtrahend).
3. "Move down" by drawing and then crossing out, the remainder from the minuend in the top row to the bottom row. The beads that are left in the bottom row is your remainder.
4. Write the numeral representing the remainder in the bottom left box.

*At the end of the problem, all of the beads in the top row should be crossed out with an "X."

20

Su M T W Th F Sa

Can you divide the beads on the table equally among your three friends? Evenly distribute the beads on each tray, starting with the largest beads. Draw your beads like you have been shown before, using cubes, squares, lines, and dots. Cross out the beads on the table with an "X" after you have placed them on your friends' trays. BONUS: Write the quotient! (How many did each friend get?)

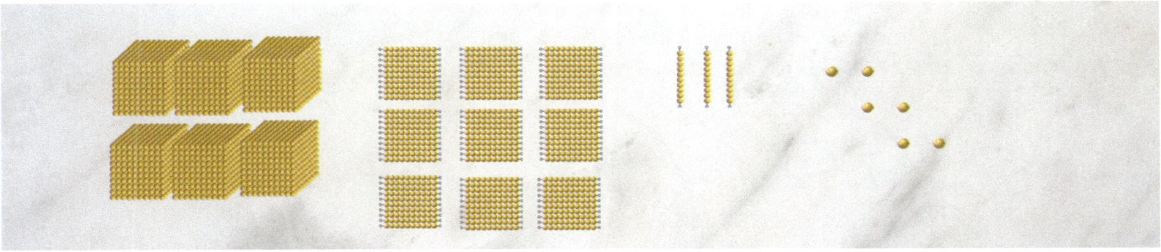

## 21

Su M T W Th F Sa

A Golden Beads Multiplication operation is shown inside the golden rectangle. Write the equation in the equation box, including the multiplicand, the multiplier, and the product.

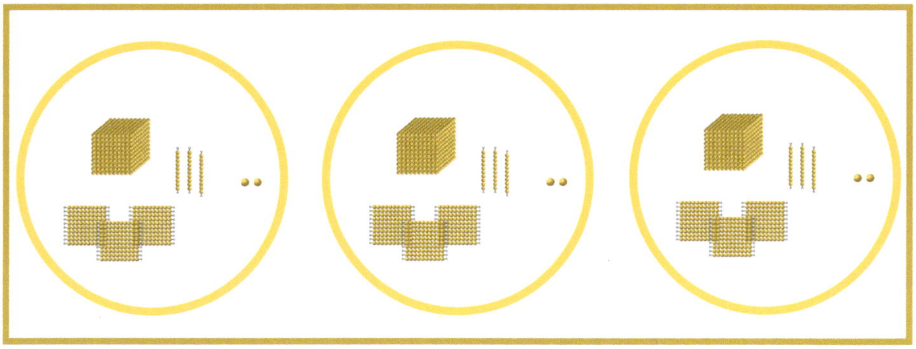

Equation Box

×

## 22

Su M T W Th F Sa

Find the matching colors of number pairs. Add them using the Strip Board. Color the first addend strip blue and the second red. Then find the sum.

3 7 9 6 5 4 2 6 8 9

### Addition Strip Board

1 2 3 4 5 6 7 8 9 10 11 12 13 14 15 16 17 18

__ + __ = ___

__ + __ = ___

__ + __ = ___

__ + __ = ___

__ + __ = ___

## 23

Su M T W Th F Sa

Draw a line from the hook to the fish with the duplicate equation. Write the answer to the equation underneath each fish. Use the strip board or a finger chart as needed.

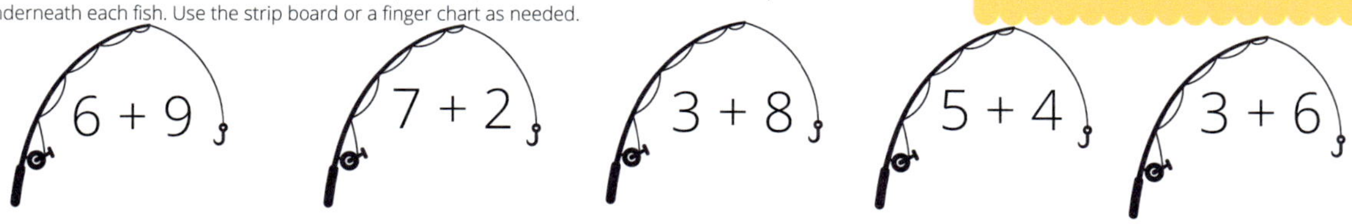


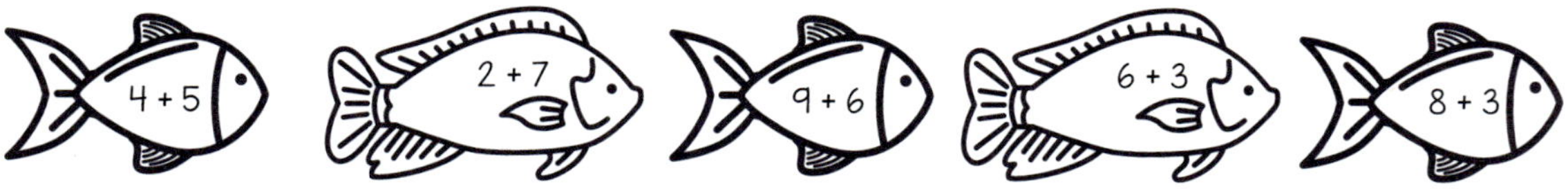

## 24

Su M T W Th F Sa

Use the Addition Strip Board below or one of the Addition Working Charts to solve each equation, if needed. Write your answers in the grid. Check your answers with Addition Chart 1 or 2.

| 1 | 2 | 3 | 4 | 5 | 6 | 7 | 8 | 9 | 10 | 11 | 12 | 13 | 14 | 15 | 16 | 17 | 18 |
|---|---|---|---|---|---|---|---|---|---|---|---|---|---|---|---|---|---|
| | | | | | | | | | | | | | | | | | |

| | | | | | | | | | | | | | |
|---|---|---|---|---|---|---|---|---|---|---|---|---|---|
| 1 | + | 1 | = | | | | 6 | + | 1 | = | | |
| 2 | + | 1 | = | | | | 7 | + | 1 | = | | |
| 3 | + | 1 | = | | | | 8 | + | 1 | = | | |
| 4 | + | 1 | = | | | | 9 | + | 1 | = | | |
| 5 | + | 1 | = | | | | | | | | | |

## 25

Su M T W Th F Sa

Solve each equation, using the Addition Strip Board below or one of the Addition Working Charts, if needed. Write your answers in the grid. Check your answers with the Addition Control Chart 1 or 2.

| 1 | 2 | 3 | 4 | 5 | 6 | 7 | 8 | 9 | 10 | 11 | 12 | 13 | 14 | 15 | 16 | 17 | 18 |
|---|---|---|---|---|---|---|---|---|---|---|---|---|---|---|---|---|---|
| | | | | | | | | | | | | | | | | | |

| | | | | | | | | | | | | | |
|---|---|---|---|---|---|---|---|---|---|---|---|---|---|
| 1 | + | 0 | = | | | | 6 | + | 0 | = | | |
| 2 | + | 0 | = | | | | 7 | + | 0 | = | | |
| 3 | + | 0 | = | | | | 8 | + | 0 | = | | |
| 4 | + | 0 | = | | | | 9 | + | 0 | = | | |
| 5 | + | 0 | = | | | | | | | | | |

## 26

Su M T W Th F Sa

Using a matching colored pencil, circle the sum on the **Addition Working Chart 3**. Then write the sum.

Check answers using **Addition Control Chart 1 or 2.**

| 0 | 1 | 2 | 3 | 4 | 5 | 6 | 7 | 8 | 9 |
|---|---|---|---|---|---|---|---|---|---|
| 1 | 2 | 3 | 4 | 5 | 6 | 7 | 8 | 9 | 10 |
| 2 | 3 | 4 | 5 | 6 | 7 | 8 | 9 | 10 | 11 |
| 3 | 4 | 5 | 6 | 7 | 8 | 9 | 10 | 11 | 12 |
| 4 | 5 | 6 | 7 | 8 | 9 | 10 | 11 | 12 | 13 |
| 5 | 6 | 7 | 8 | 9 | 10 | 11 | 12 | 13 | 14 |
| 6 | 7 | 8 | 9 | 10 | 11 | 12 | 13 | 14 | 15 |
| 7 | 8 | 9 | 10 | 11 | 12 | 13 | 14 | 15 | 16 |
| 8 | 9 | 10 | 11 | 12 | 13 | 14 | 15 | 16 | 17 |
| 9 | 10 | 11 | 12 | 13 | 14 | 15 | 16 | 17 | 18 |

| | | | | | |
|---|---|---|---|---|---|
| 4 | + | 8 | | | |
| 2 | + | =9 | = | | |
| 6 | + | 7 | = | | |
| 5 | + | 6 | = | | |
| 8 | + | 8 | = | | |

## 27

Su M T W Th F Sa

Using a matching colored pencil, circle the sum on the **Addition Working Chart 4.** Then write the sum.

Check answers using **Addition Control Chart 1 or 2.**

| | | | | | | | | | |
|---|---|---|---|---|---|---|---|---|---|
| 1 | 2 | | | | | | | | |
| 2 | 3 | 4 | | | | | | | |
| 3 | 4 | 5 | 6 | | | | | | |
| 4 | 5 | 6 | 7 | 8 | | | | | |
| 5 | 6 | 7 | 8 | 9 | 10 | | | | |
| 6 | 7 | 8 | 9 | 10 | 11 | 12 | | | |
| 7 | 8 | 9 | 10 | 11 | 12 | 13 | 14 | | |
| 8 | 9 | 10 | 11 | 12 | 13 | 14 | 15 | 16 | |
| 9 | 10 | 11 | 12 | 13 | 14 | 15 | 16 | 17 | 18 |

| | | | | | |
|---|---|---|---|---|---|
| 7 | + | 7 | = | | |
| 4 | + | 6 | = | | |
| 3 | + | 9 | = | | |
| 6 | + | 8 | = | | |
| 8 | + | 9 | = | | |

## 28

Su M T W Th F Sa

Using a matching colored pencil, circle the sum on the **Addition Working Chart 5**. Then write the sum.

Check answers using **Addition Control Chart 1 or 2.**

| | | | | | | | | | |
|---|---|---|---|---|---|---|---|---|---|
| 1 | 2 | | | | | | | | |
| 2 | 3 | 4 | | | | | | | |
| 3 | | 5 | 6 | | | | | | |
| 4 | | | 7 | 8 | | | | | |
| 5 | | | | 9 | 10 | | | | |
| 6 | | | | | 11 | 12 | | | |
| 7 | | | | | | 13 | 14 | | |
| 8 | | | | | | | 15 | 16 | |
| 9 | | | | | | | | 17 | 18 |

| | | | | | |
|---|---|---|---|---|---|
| 1 | + | 8 | = | | |
| 7 | + | 4 | = | | |
| 5 | + | 9 | = | | |
| 6 | + | 3 | = | | |
| 8 | + | 7 | = | | |

**29**

Su M T W Th F Sa

Finish the series of subtraction equations using the **Subtraction Strip Board.**
Color in the number strips then write the answers on the lines.

9 - 1 = 8

9 - 2 = ____

9 - 3 = ____

9 - 4 = ____

9 - 5 = ____

9 - 6 = ____

9 - 7 = ____

9 - 8 = ____

9 - 9 = ____

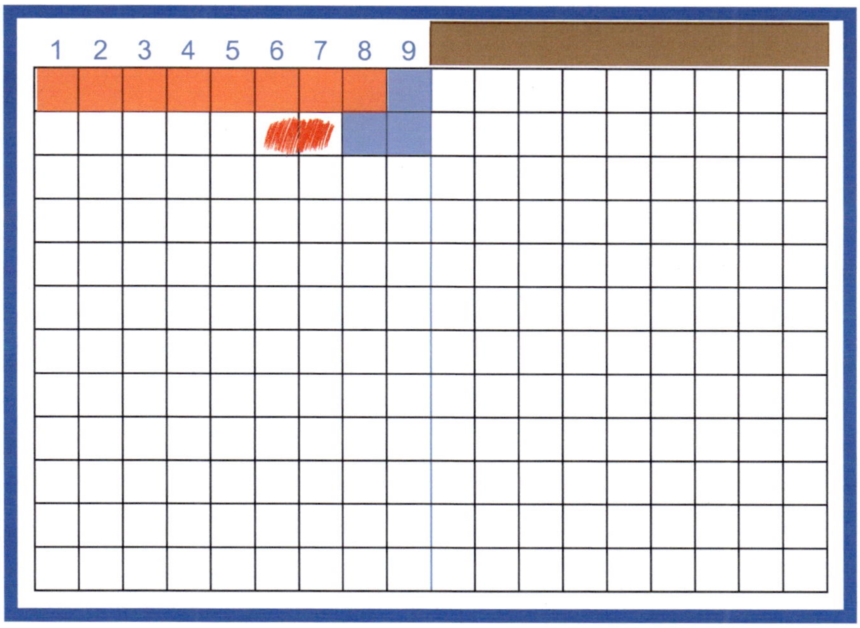

**30**

Su M T W Th F Sa

Draw the bead bars for the **House of Five.**

Write out three equations for the sum of 5, using the bead house.

## 31

Su M T W Th F Sa

Write the numerals and trace the number words.

10 ten

twenty

thirty

forty

fifty

sixty

## 32

Su M T W Th F Sa

Write the correct symbol in the circle.
Greater than >
Less than <

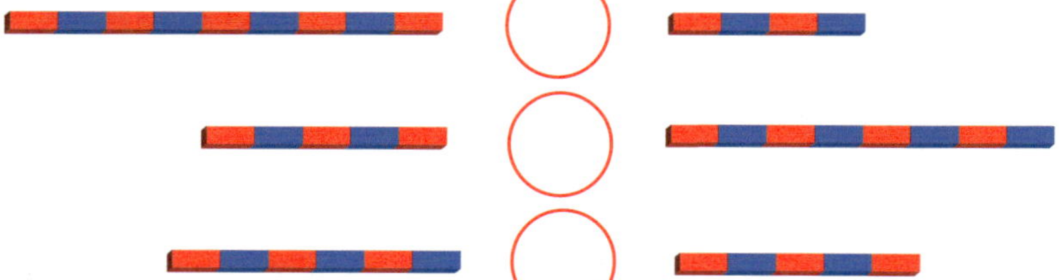

16 ◯ 26

8 ◯ 18

Hint: The alligator eats the greater number. We call him the "greater gator."

## 33

Su M T W Th F Sa

Draw the rod that you need to make ten. Then write the addition equation.

10

___ + ___ = ___

___ + ___ = ___

___ + ___ = ___

## 34

Su M T W Th F Sa

**Background:**
The only place value that needs to be examined to determine if a number is even or odd is the units place.
If a number has a 0, 2, 4, 6, or 8 in the units place, the number is *even*.
If a number has a 1, 3, 5, 7, or 9 in the units place, the number is *odd*.
Remember, if the units can be *paired*, the number is even.
If the units have an "odd one out" by itself, the number is odd.

**Directions:**
Draw the dots in pairs to represent only the *units* in the following numbers. Then circle the word "even" or "odd."

hint: draw 10 counters

| 23 | 140 | 5734 |
|---|---|---|
| even odd | even odd | even odd |

## 35

Su M T W Th F Sa

# Make 10 Snake Game

Directions:
1. Color the bead bars in the snake their proper colors. (1=red, 2=green, 3=pink, 4=yellow, 5=turquoise, 6=lavender, 7=white, 8=brown, 9=dark blue).
2. Count ten beads. Put a pencil mark where you got to 10. Color a golden bead ten-bar at the bottom of the box and cross out the ten beads before your pencil mark. This shows an "exchange."
3. Continue counting to 10, putting a pencil mark, coloring the golden bead ten-bar, and crossing out the beads.
4. When you get to the end, you may need to exchange any beads less than ten for a black-and-white bead bar. Draw this black-and-white bead bar on the tail of the golden snake according to the key.

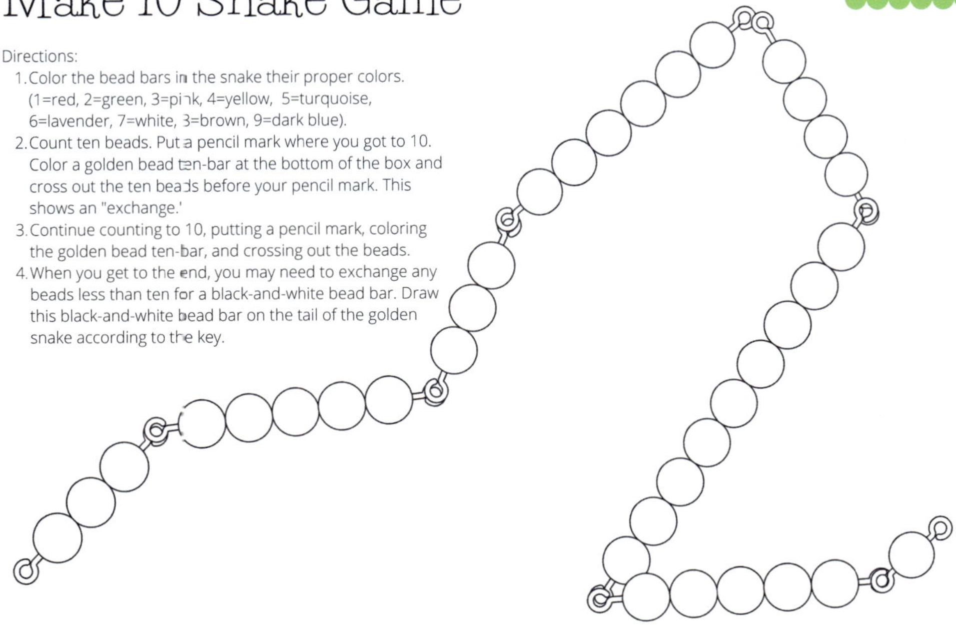

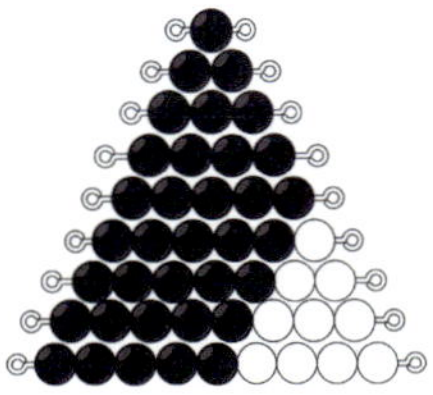

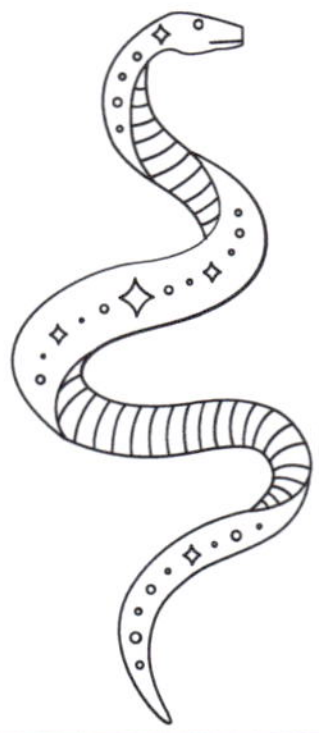

## 36

Su M T W Th F Sa

Color the Short Bead Bars, then write the teen number made.

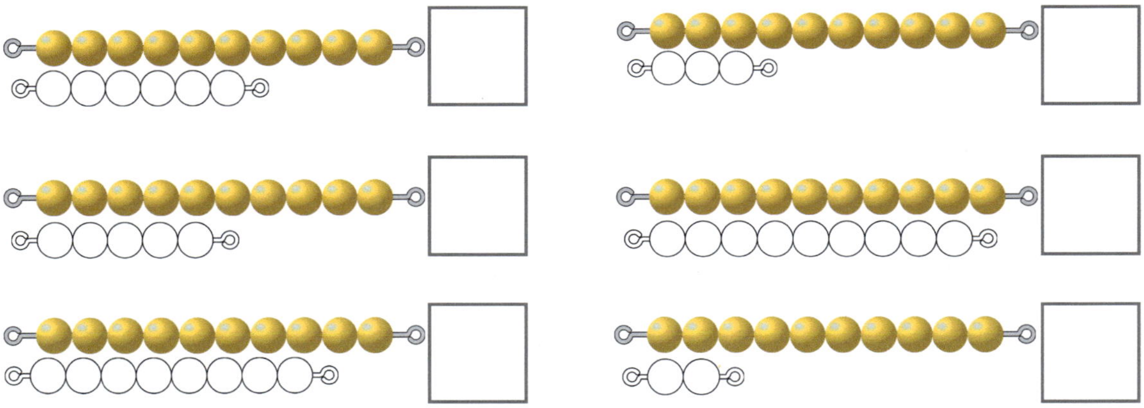

## 37

Su M T W Th F Sa

1. Draw a line from the numeral tile to the space it belongs on the teen board for numbers 16-19.
2. Write in the numerals needed to make 16-19 on the other board.
3. Trace or copy the words for the teen numbers.

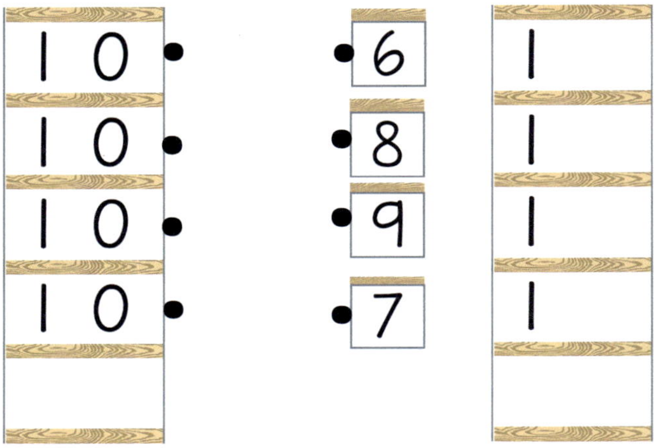


sixteen

seventeen

eighteen

nineteen

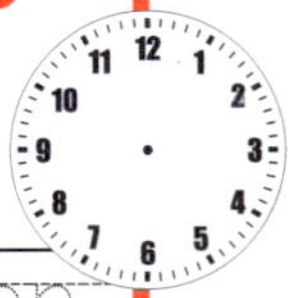

## 38

Su M T W Th F Sa

What teen value is made from combining the two number rods? Write the teen numeral in the box.

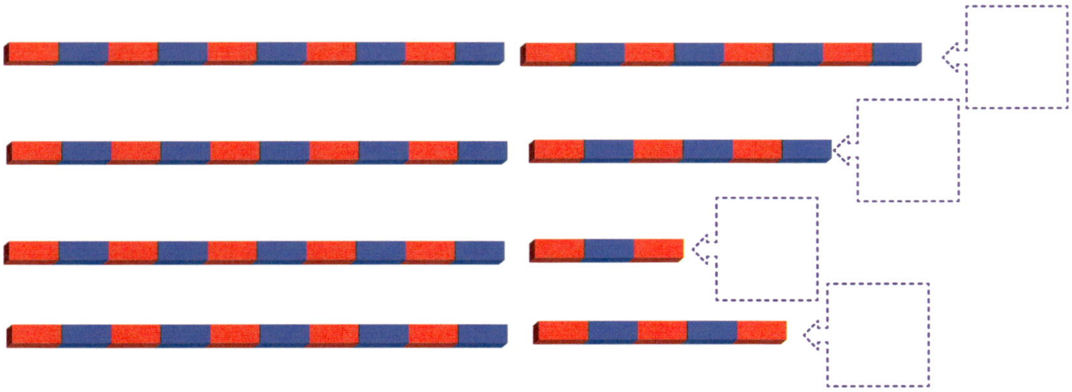

## 39

Su M T W Th F Sa

Write the numbers represented by the beads onto the ten boards.

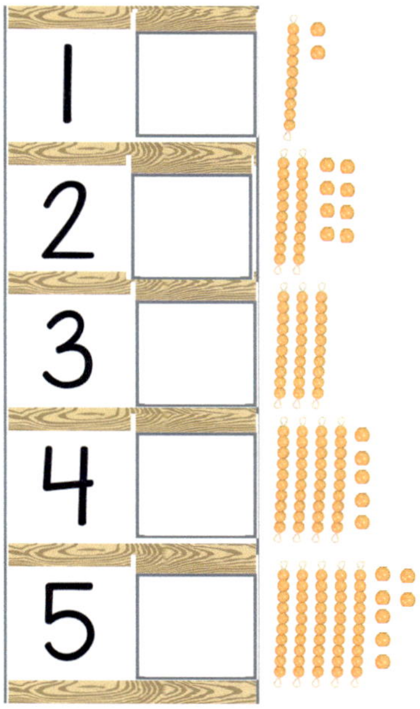


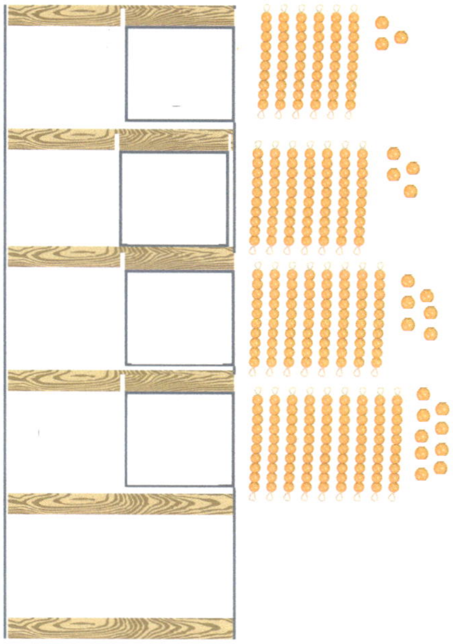

## 40

Su M T W Th F Sa

Color the beads along the 100 Chain according to the key below the chain.
*The first one is colored for you.*

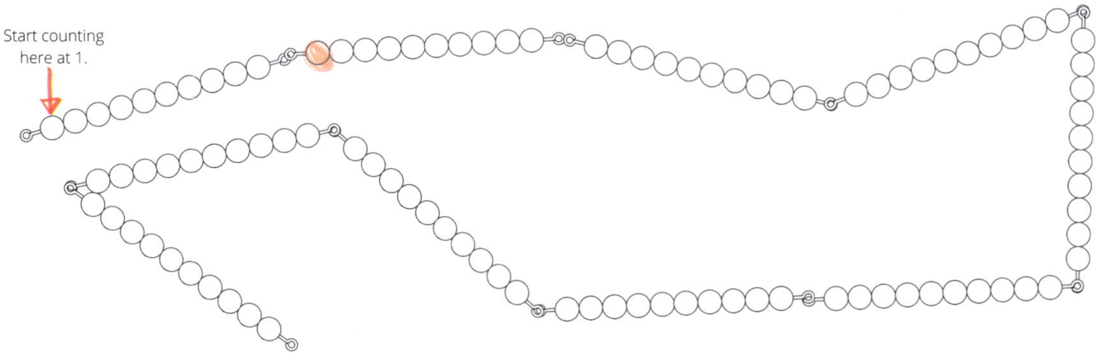

**11 = red** **30 = blue** **57 = green** **75 = yellow** **92 = pink**

# Looping Review Exercises

## 41

Su M T W Th F Sa

Write the number on the line that belongs at each location on the blank Hundred Board

A ______
B ______
C ______
D______
E ______
F ______
G ______

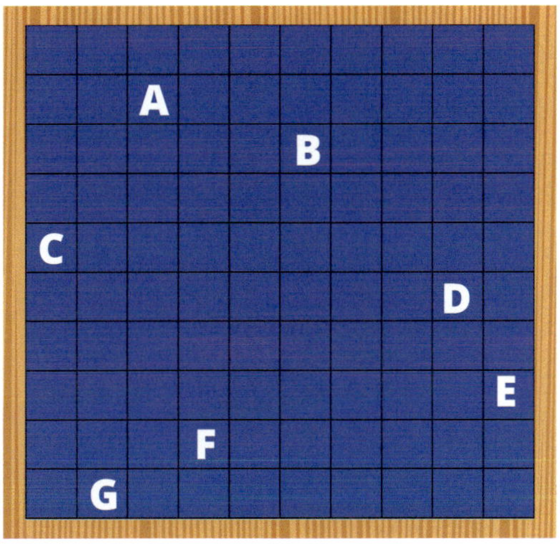


Check your work against the Hundred Board in Looping Review Exercise #43 on this page. Put a check in the box when you finish checking your work. .

I checked my work!

## 12

Su M T W Th F Sa

Fill in the numbers of the beads on the squaring chain. Count the beads in the square.

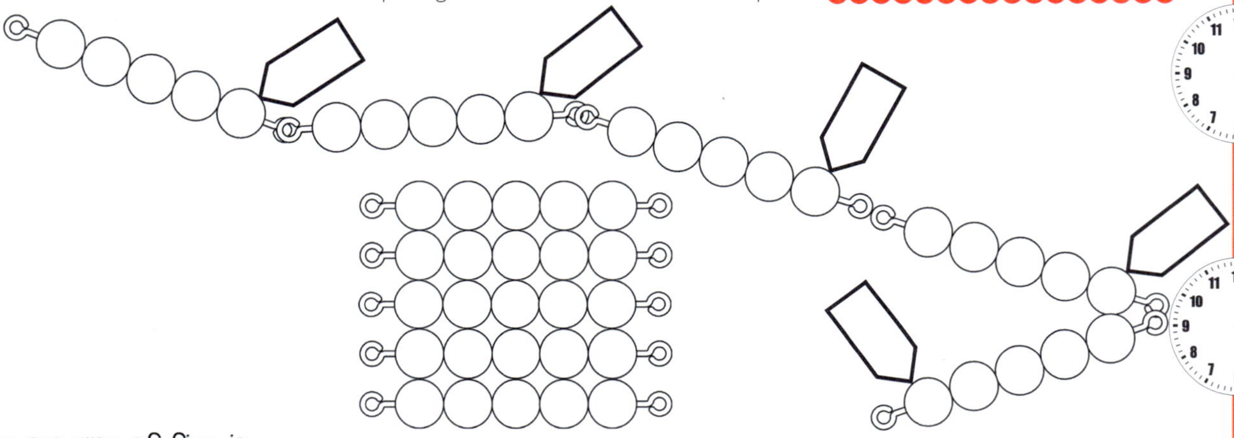

The square of five is______ .

## 43

Su M T W Th F Sa

Highlight the multiples of ten. (Skip count by 10's.)
Circle the multiples of five. (Skip count by 5's.)

| | | | | | | | | | |
|---|---|---|---|---|---|---|---|---|---|
| 1 | 2 | 3 | 4 | 5 | 6 | 7 | 8 | 9 | 10 |
| 11 | 12 | 13 | 14 | 15 | 16 | 17 | 18 | 19 | 20 |
| 21 | 22 | 23 | 24 | 25 | 26 | 27 | 28 | 29 | 30 |
| 31 | 32 | 33 | 34 | 35 | 36 | 37 | 38 | 39 | 40 |
| 41 | 42 | 43 | 44 | 45 | 46 | 47 | 48 | 49 | 50 |
| 51 | 52 | 53 | 54 | 55 | 56 | 57 | 58 | 59 | 60 |
| 61 | 62 | 63 | 64 | 65 | 66 | 67 | 68 | 69 | 70 |
| 71 | 72 | 73 | 74 | 75 | 76 | 77 | 78 | 79 | 80 |
| 81 | 82 | 83 | 84 | 85 | 86 | 87 | 88 | 89 | 90 |
| 91 | 92 | 93 | 94 | 95 | 96 | 97 | 98 | 99 | 100 |

## 44

Su M T W Th F Sa

Count by tens. Write the numbers on the arrows.

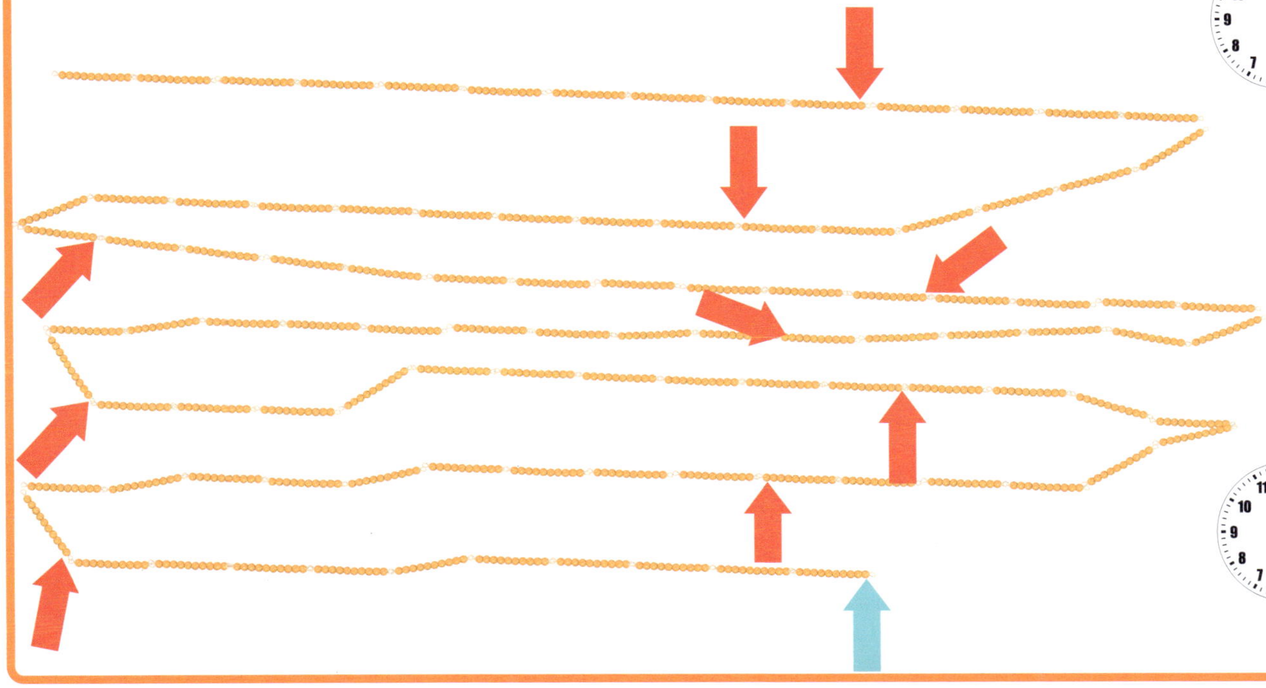

## 45

Su M T W Th F Sa

Draw a line from the quantity in the first column to the equivalent quantity in the middle column. Then draw a line from the cuantity in the middle column to the equivalent numeral in the right column.

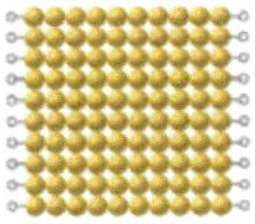

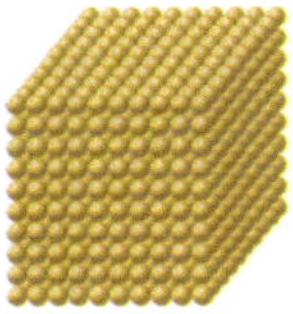

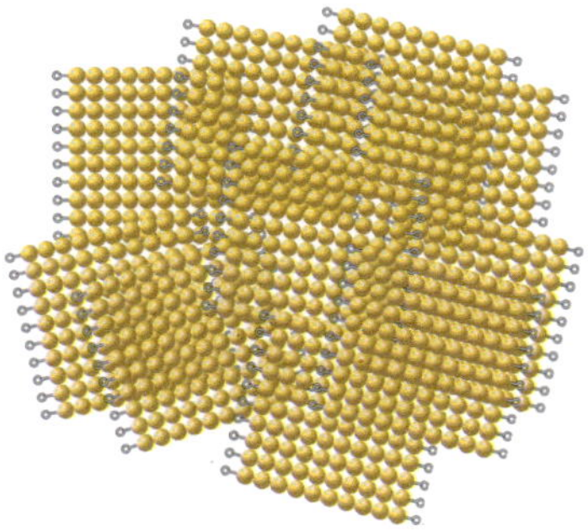

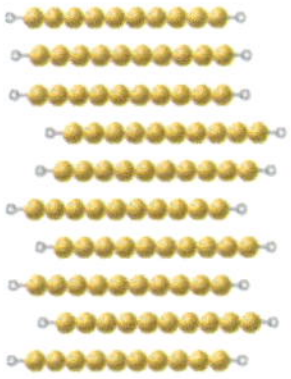

10

1000

100

# Looping Review Exercises

## 46

Su M T W Th F Sa

Color the corresponding amount of beads.

2 8 9 0

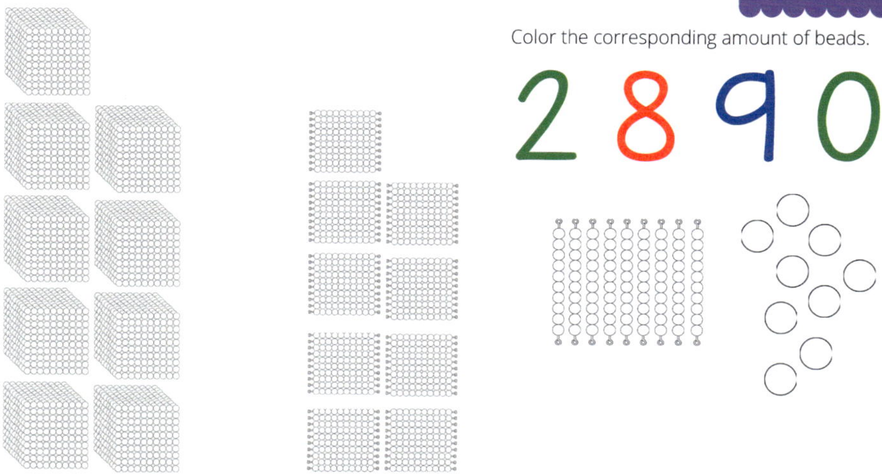

## 47

Su M T W Th F Sa

Circle or shade in the corresponding decimal cards for this amount. Then write the number.

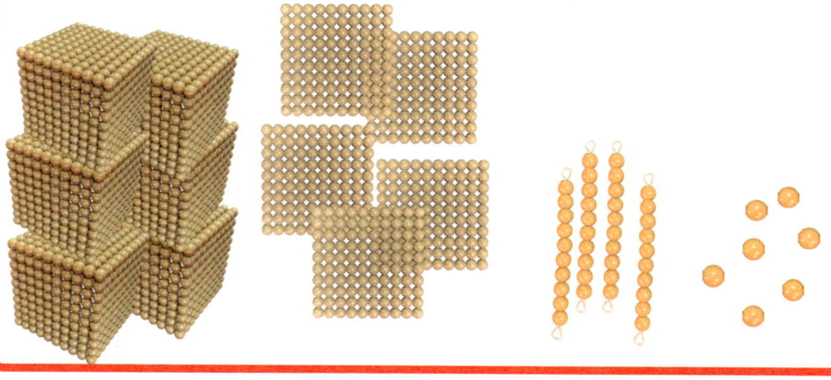

| | | | |
|---|---|---|---|
| 1000 | 100 | 10 | 1 |
| 2000 | 200 | 20 | 2 |
| 3000 | 300 | 30 | 3 |
| 4000 | 400 | 40 | 4 |
| 5000 | 500 | 50 | 5 |
| 6000 | 600 | 60 | 6 |
| 7000 | 700 | 70 | 7 |
| 8000 | 800 | 80 | 8 |
| 9000 | 900 | 90 | 9 |

## 48

Su M T W Th F Sa

Using the simple drawing method for the Golden Beads, draw out the problem on the addition mat. When you move the beads "down" under the red addition line, redraw them in the lower section and cross them out from the upper sections.Write the numeral answer in the bottom left box.

| | | | | |
|---|---|---|---|---|
| 2314 | | | | |
| + 2675 | | | | |
| | | | | |

1000 100 10 1

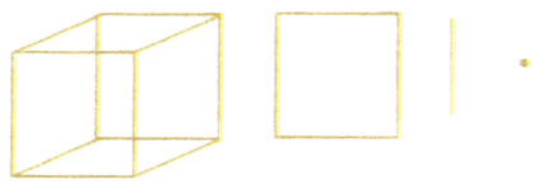

## 49

Su M T W Th F Sa

1. Write the numerals in the left column for the minuend and the subtrahend.
2. Cross out the beads from the top row (the minuend) that have been "moved down" to the middle row (the subtrahend).
3. "Move down" by drawing and then crossing out, the remainder from the minuend in the top row to the bottom row. The beads that are left in the bottom row is your remainder.
4. Write the numeral representing the remainder in the bottom left box.

*At the end of the problem, all of the beads in the top row should be crossed out with an "X."

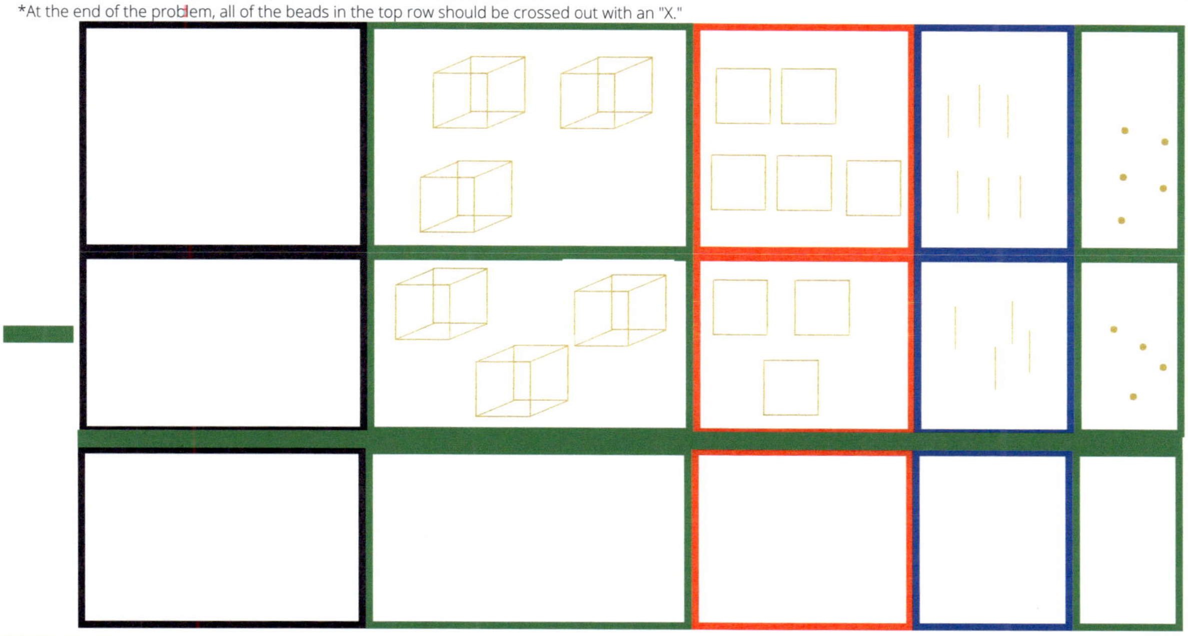

## 50

Su M T W Th F Sa

Can you divide the beads on the table equally among your three friends? Evenly distribute the beads on each tray, starting with the largest beads. Draw your beads like you have been shown before, using cubes, squares, lines, and dots. Cross out the beads on the table with an "X" after you have placed them on your friends' trays. BONUS: Write the quotient! (How many did each friend get?)

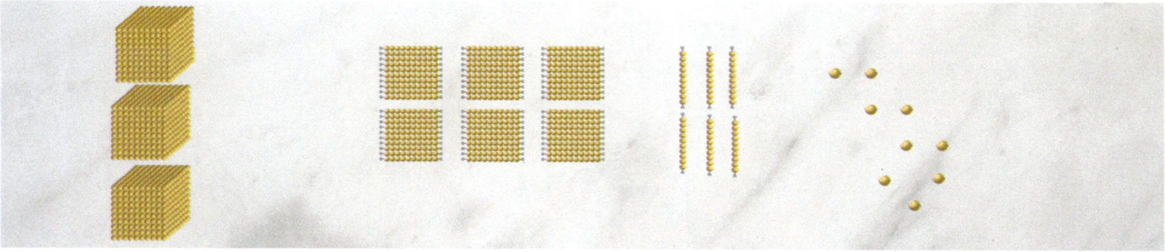

# Looping Review Exercises

## 51

Su M T W Th F Sa

A Golden Beads Multiplication operation is shown inside the golden rectangle. Write the equation in the equation box, including the multiplicand, the multiplier, and the product..

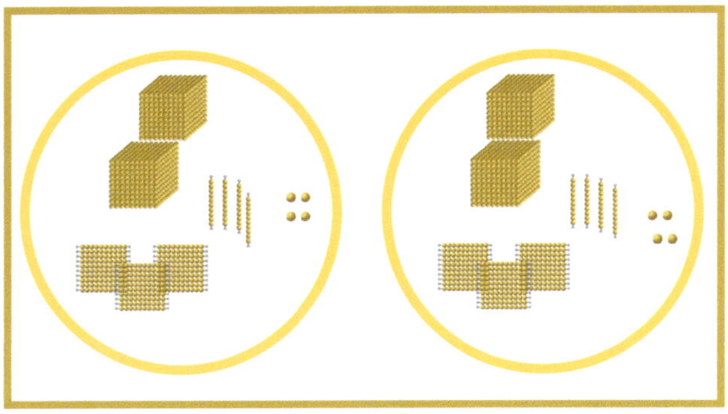

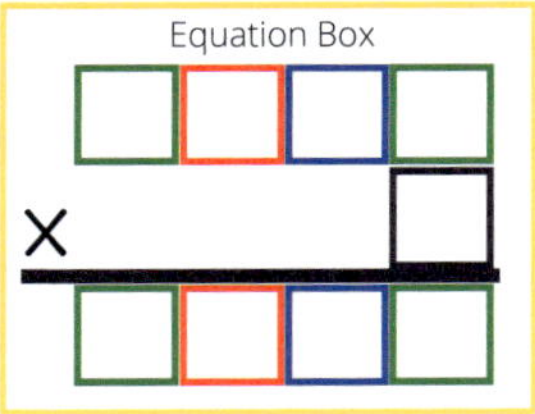

## 52

Su M T W Th F Sa

Find the matching colors of number pairs. Add them using the Strip Board. Color the first addend strip blue and the second red. Then find the sum.

3 2 1 5 9 7 4 6 7 8

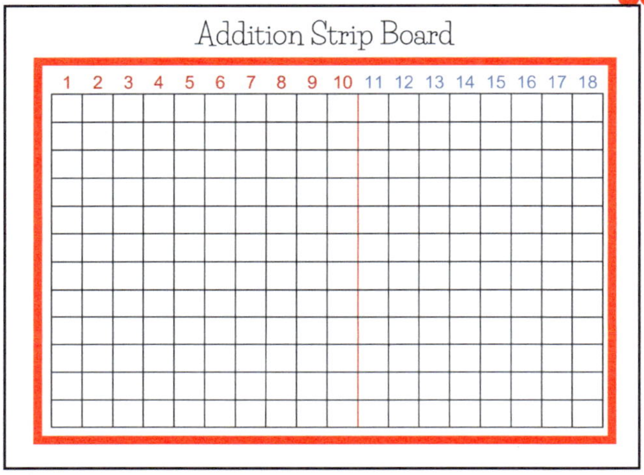


___ + ___ = ___

___ + ___ = ___

___ + ___ = ___

___ + ___ = ___

___ + ___ = ___

## 53

Su M T W Th F Sa

Draw a line from the hook to the fish with the duplicate equation. Write the answer to the equation underneath each fish. Use the strip board or a finger chart as needed.

## 54

Su M T W Th F Sa

Use the Addition Strip Board or one of the Addition Working Charts to solve each equation. Write your answers in the grid. Check your answers with Addition Chart 1.

| 1 | 2 | 3 | 4 | 5 | 6 | 7 | 8 | 9 | 10 | 11 | 12 | 13 | 14 | 15 | 16 | 17 | 18 |
|---|---|---|---|---|---|---|---|---|---|---|---|---|---|---|---|---|---|
| | | | | | | | | | | | | | | | | | |

| | | | | | | | | | | | | |
|---|---|---|---|---|---|---|---|---|---|---|---|---|
| 1 | + | 2 | = | | | | 6 | + | 2 | = | | |
| 2 | + | 2 | = | | | | 7 | + | 2 | = | | |
| 3 | + | 2 | = | | | | 8 | + | 2 | = | | |
| 4 | + | 2 | = | | | | 9 | + | 2 | = | | |
| 5 | + | 2 | = | | | | | | | | | |

## 55

Su M T W Th F Sa

Solve each equation, using the Addition Strip Board or one of the Addition Working Charts, if needed. Write your answers in the grid.

| 1 | 2 | 3 | 4 | 5 | 6 | 7 | 8 | 9 | 10 | 11 | 12 | 13 | 14 | 15 | 16 | 17 | 18 |
|---|---|---|---|---|---|---|---|---|---|---|---|---|---|---|---|---|---|
| | | | | | | | | | | | | | | | | | |

| | | | | | | | | | | | | |
|---|---|---|---|---|---|---|---|---|---|---|---|---|
| 1 | + | 7 | = | | | | 6 | + | 2 | = | | |
| 2 | + | 5 | = | | | | 7 | + | 1 | = | | |
| 3 | + | 9 | = | | | | 8 | + | 3 | = | | |
| 4 | + | 4 | = | | | | 9 | + | 6 | = | | |
| 5 | + | 8 | = | | | | | | | | | |

## 56

Su M T W Th F Sa

Using a matching colored pencil, circle the sum on the **Addition Working Chart 3**. Then write the sum.

| 0 | 1 | 2 | 3 | 4 | 5 | 6 | 7 | 8 | 9 |
|---|---|---|---|---|---|---|---|---|---|
| 1 | 2 | 3 | 4 | 5 | 6 | 7 | 8 | 9 | 10 |
| 2 | 3 | 4 | 5 | 6 | 7 | 8 | 9 | 10 | 11 |
| 3 | 4 | 5 | 6 | 7 | 8 | 9 | 10 | 11 | 12 |
| 4 | 5 | 6 | 7 | 8 | 9 | 10 | 11 | 12 | 13 |
| 5 | 6 | 7 | 8 | 9 | 10 | 11 | 12 | 13 | 14 |
| 6 | 7 | 8 | 9 | 10 | 11 | 12 | 13 | 14 | 15 |
| 7 | 8 | 9 | 10 | 11 | 12 | 13 | 14 | 15 | 16 |
| 8 | 9 | 10 | 11 | 12 | 13 | 14 | 15 | 16 | 17 |
| 9 | 10 | 11 | 12 | 13 | 14 | 15 | 16 | 17 | 18 |

6 + 5

3 + = 7 =

1 + 4 =

8 + 3 =

9 + 6 =

## 57

Su M T W Th F Sa

Using a matching colored pencil, circle the sum on the **Addition Working Chart 4.** Then write the sum.

| | | | | | | | | | |
|---|---|---|---|---|---|---|---|---|---|
| 1 | 2 | | | | | | | | |
| 2 | 3 | 4 | | | | | | | |
| 3 | 4 | 5 | 6 | | | | | | |
| 4 | 5 | 6 | 7 | 8 | | | | | |
| 5 | 6 | 7 | 8 | 9 | 10 | | | | |
| 6 | 7 | 8 | 9 | 10 | 11 | 12 | | | |
| 7 | 8 | 9 | 10 | 11 | 12 | 13 | 14 | | |
| 8 | 9 | 10 | 11 | 12 | 13 | 14 | 15 | 16 | |
| 9 | 10 | 11 | 12 | 13 | 14 | 15 | 16 | 17 | 18 |

6 + 6 =

5 + 8 =

2 + 4 =

7 + 6 =

3 + 4 =

## 58

Su M T W Th F Sa

Using a matching colored pencil, circle the sum on the **Addition Working Chart 5**. Then write the sum.

| | | | | | | | | | |
|---|---|---|---|---|---|---|---|---|---|
| 1 | 2 | | | | | | | | |
| 2 | 3 | 4 | | | | | | | |
| 3 | | 5 | 6 | | | | | | |
| 4 | | | 7 | 8 | | | | | |
| 5 | | | | 9 | 10 | | | | |
| 6 | | | | | 11 | 12 | | | |
| 7 | | | | | | 13 | 14 | | |
| 8 | | | | | | | 15 | 16 | |
| 9 | | | | | | | | 17 | 18 |

2 + 8 =

5 + 3 =

6 + 2 =

1 + 8 =

9 + 7 =

**59**

Su M T W Th F Sa

Finish the series of subtraction equations using the **Subtraction Strip Board.** Color in the number strips then write the answers on the lines. The first equation is colored for you.

6 - 1 = ____

6 - 2 = ____

6 - 3 = ____

6 - 4 = ____

6 - 5 = ____

6 - 6 = ____

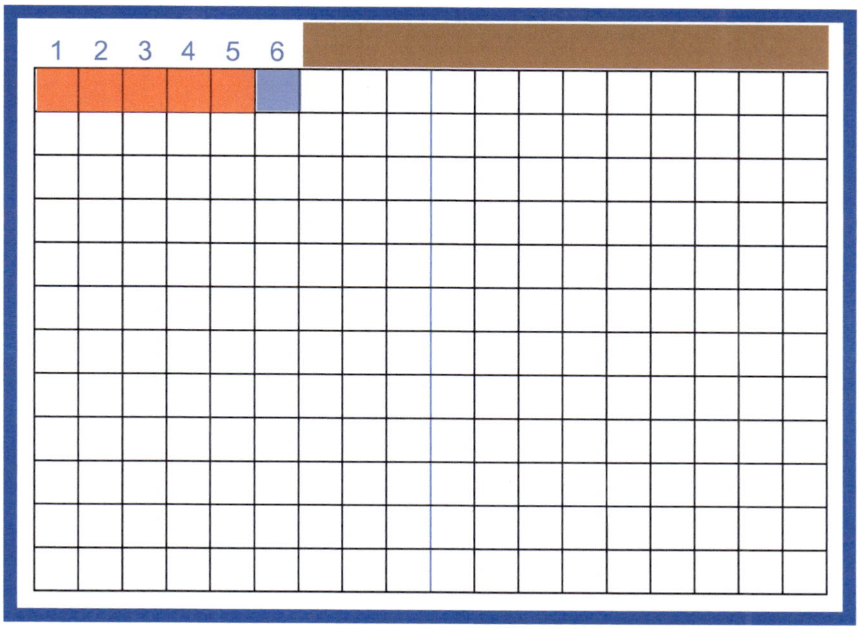

**60**

Su M T W Th F Sa

Draw the bead bars for this house. Write the corresponding addition equations.

____________________

____________________

____________________

____________________

____________________

# Appendix

For extra durability, laminate the charts and manipulatives in the Appendix. You may not want to laminate the stamps for the stamp game, though, as the corners and edges can be sharp.

Store the materials alongside the Workbook in a folder, in page protectors in a binder, or however you decide. Keep in mind portability and durability. Zip-top clear bags may not be fancy, but they work well to hold pieces!

# Addition Strip Board

| 1 | 2 | 3 | 4 | 5 | 6 | 7 | 8 | 9 | 10 | 11 | 12 | 13 | 14 | 15 | 16 | 17 | 18 |
|---|---|---|---|---|---|---|---|---|---|---|---|---|---|---|---|---|---|
| | | | | | | | | | | | | | | | | | |
| | | | | | | | | | | | | | | | | | |
| | | | | | | | | | | | | | | | | | |
| | | | | | | | | | | | | | | | | | |
| | | | | | | | | | | | | | | | | | |
| | | | | | | | | | | | | | | | | | |
| | | | | | | | | | | | | | | | | | |
| | | | | | | | | | | | | | | | | | |
| | | | | | | | | | | | | | | | | | |
| | | | | | | | | | | | | | | | | | |
| | | | | | | | | | | | | | | | | | |
| | | | | | | | | | | | | | | | | | |

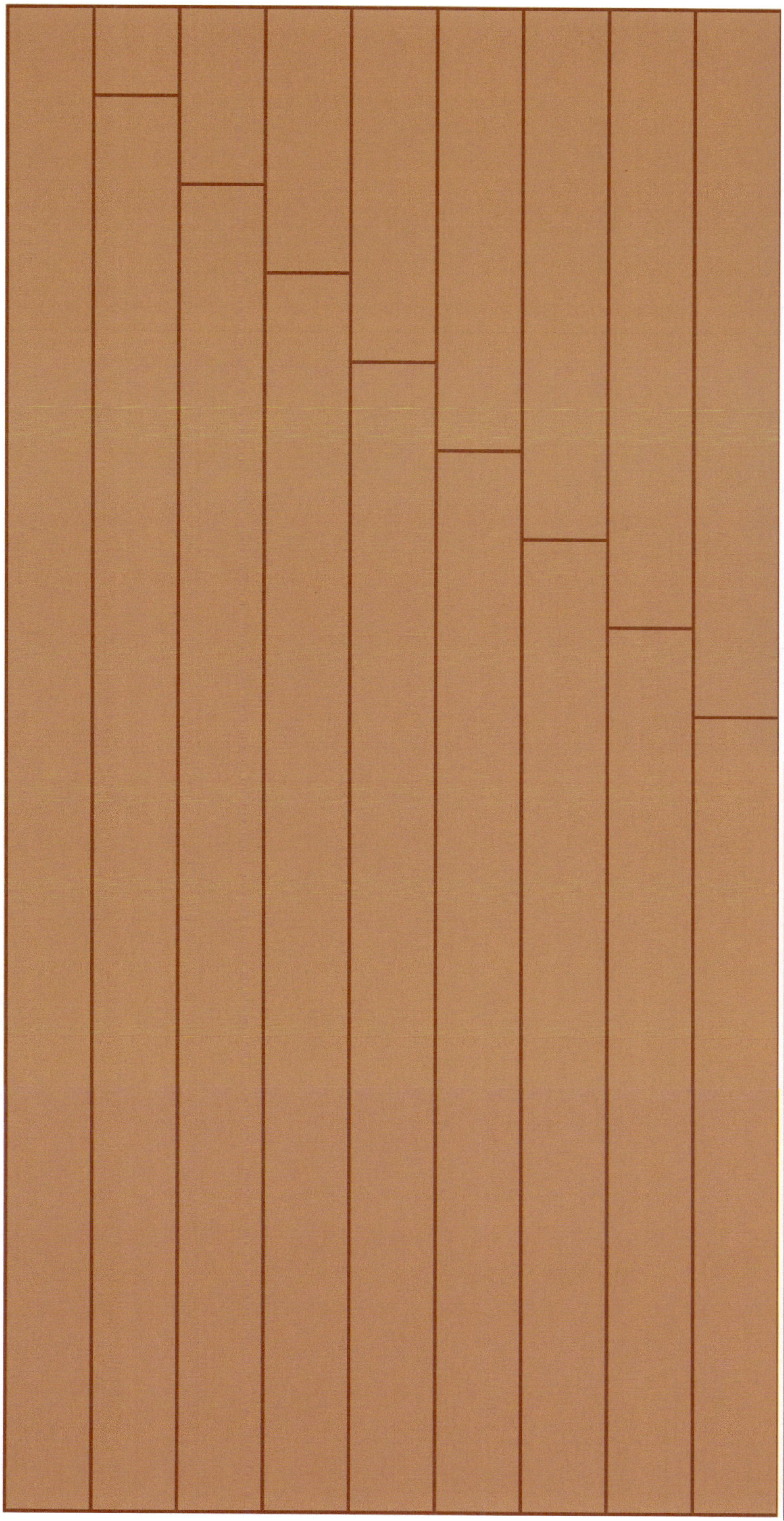

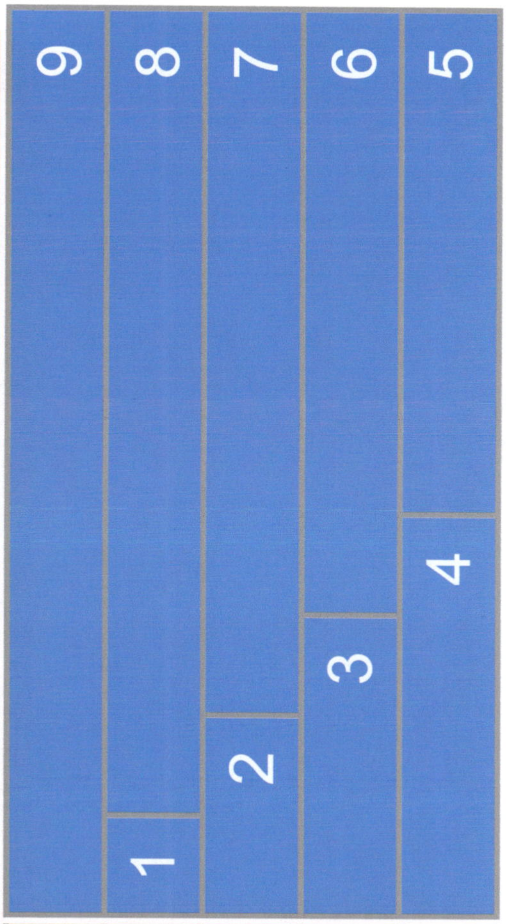
9
8
7
6
5
1
2
3
4

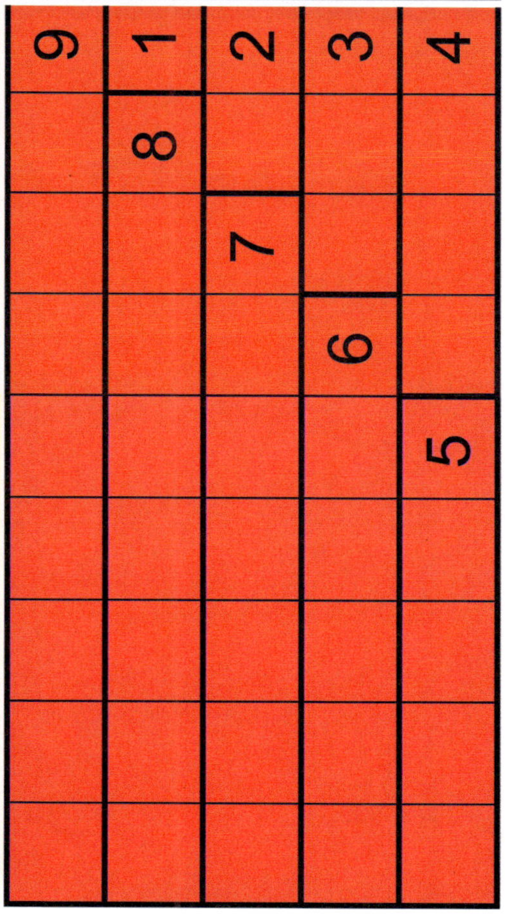
9
1
2
3
4
8
7
6
5

A-4

# Addition Control Chart 1

| | | | | | | | | |
|---|---|---|---|---|---|---|---|---|
| 1 + 1 = 2 | 2 + 1 = 3 | 3 + 1 = 4 | 4 + 1 = 5 | 5 + 1 = 6 | 6 + 1 = 7 | 7 + 1 = 8 | 8 + 1 = 9 | 9 + 1 = 10 |
| 1 + 2 = 3 | 2 + 2 = 4 | 3 + 2 = 5 | 4 + 2 = 6 | 5 + 2 = 7 | 6 + 2 = 8 | 7 + 2 = 9 | 8 + 2 = 10 | 9 + 2 = 11 |
| 1 + 3 = 4 | 2 + 3 = 5 | 3 + 3 = 6 | 4 + 3 = 7 | 5 + 3 = 8 | 6 + 3 = 9 | 7 + 3 = 10 | 8 + 3 = 11 | 9 + 3 = 12 |
| 1 + 4 = 5 | 2 + 4 = 6 | 3 + 4 = 7 | 4 + 4 = 8 | 5 + 4 = 9 | 6 + 4 = 10 | 7 + 4 = 11 | 8 + 4 = 12 | 9 + 4 = 13 |
| 1 + 5 = 6 | 2 + 5 = 7 | 3 + 5 = 8 | 4 + 5 = 9 | 5 + 5 = 10 | 6 + 5 = 11 | 7 + 5 = 12 | 8 + 5 = 13 | 9 + 5 = 14 |
| 1 + 6 = 7 | 2 + 6 = 8 | 3 + 6 = 9 | 4 + 6 = 10 | 5 + 6 = 11 | 6 + 6 = 12 | 7 + 6 = 13 | 8 + 6 = 14 | 9 + 6 = 15 |
| 1 + 7 = 8 | 2 + 7 = 9 | 3 + 7 = 10 | 4 + 7 = 11 | 5 + 7 = 12 | 6 + 7 = 13 | 7 + 7 = 14 | 8 + 7 = 15 | 9 + 7 = 16 |
| 1 + 8 = 9 | 2 + 8 = 10 | 3 + 8 = 11 | 4 + 8 = 12 | 5 + 8 = 13 | 6 + 8 = 14 | 7 + 8 = 15 | 8 + 8 = 16 | 9 + 8 = 17 |
| 1 + 9 = 10 | 2 + 9 = 11 | 3 + 9 = 12 | 4 + 9 = 13 | 5 + 9 = 14 | 6 + 9 = 15 | 7 + 9 = 16 | 8 + 9 = 17 | 9 + 9 = 18 |

# Addition Control Chart 2

| | | | | | | | | |
|---|---|---|---|---|---|---|---|---|
| 1 + 1 = 2 | | | | | | | | |
| 1 + 2 = 3 | | | | | | | | |
| 1 + 3 = 4 | 2 + 2 = 4 | | | | | | | |
| 1 + 4 = 5 | 2 + 3 = 5 | | | | | | | |
| 1 + 5 = 6 | 2 + 4 = 6 | 3 + 3 = 6 | | | | | | |
| 1 + 6 = 7 | 2 + 5 = 7 | 3 + 4 = 7 | | | | | | |
| 1 + 7 = 8 | 2 + 6 = 8 | 3 + 5 = 8 | 4 + 4 = 8 | | | | | |
| 1 + 8 = 9 | 2 + 7 = 9 | 3 + 6 = 9 | 4 + 5 = 9 | | | | | |
| 1 + 9 = 10 | 2 + 8 = 10 | 3 + 7 = 10 | 4 + 6 = 10 | 5 + 5 = 10 | | | | |
| | 2 + 9 = 11 | 3 + 8 = 11 | 4 + 7 = 11 | 5 + 6 = 11 | | | | |
| | | 3 + 9 = 12 | 4 + 8 = 12 | 5 + 7 = 12 | 6 + 6 = 12 | | | |
| | | | 4 + 9 = 13 | 5 + 8 = 13 | 6 + 7 = 13 | | | |
| | | | | 5 + 9 = 14 | 6 + 8 = 14 | 7 + 7 = 14 | | |
| | | | | | 6 + 9 = 15 | 7 + 8 = 15 | | |
| | | | | | | 7 + 9 = 16 | 8 + 8 = 16 | |
| | | | | | | | 8 + 9 = 17 | |
| | | | | | | | | 9 + 9 = 18 |

# Addition Working Chart 3

| 0 | 1 | 2 | 3 | 4 | 5 | 6 | 7 | 8 | 9 |
|---|---|---|---|---|---|---|---|---|---|
| 1 | 2 | 3 | 4 | 5 | 6 | 7 | 8 | 9 | 10 |
| 2 | 3 | 4 | 5 | 6 | 7 | 8 | 9 | 10 | 11 |
| 3 | 4 | 5 | 6 | 7 | 8 | 9 | 10 | 11 | 12 |
| 4 | 5 | 6 | 7 | 8 | 9 | 10 | 11 | 12 | 13 |
| 5 | 6 | 7 | 8 | 9 | 10 | 11 | 12 | 13 | 14 |
| 6 | 7 | 8 | 9 | 10 | 11 | 12 | 13 | 14 | 15 |
| 7 | 8 | 9 | 10 | 11 | 12 | 13 | 14 | 15 | 16 |
| 8 | 9 | 10 | 11 | 12 | 13 | 14 | 15 | 16 | 17 |
| 9 | 10 | 11 | 12 | 13 | 14 | 15 | 16 | 17 | 18 |

# Addition Working Chart 4

| | | | | | | | | | |
|---|---|---|---|---|---|---|---|---|---|
| 1 | 2 | | | | | | | | |
| 2 | 3 | 4 | | | | | | | |
| 3 | 4 | 5 | 6 | | | | | | |
| 4 | 5 | 6 | 7 | 8 | | | | | |
| 5 | 6 | 7 | 8 | 9 | 10 | | | | |
| 6 | 7 | 8 | 9 | 10 | 11 | 12 | | | |
| 7 | 8 | 9 | 10 | 11 | 12 | 13 | 14 | | |
| 8 | 9 | 10 | 11 | 12 | 13 | 14 | 15 | 16 | |
| 9 | 10 | 11 | 12 | 13 | 14 | 15 | 16 | 17 | 18 |

# Addition Working Chart 5

| | | | | | | | | | |
|---|---|---|---|---|---|---|---|---|---|
| 1 | 2 | | | | | | | | |
| 2 | 3 | 4 | | | | | | | |
| 3 | | 5 | 6 | | | | | | |
| 4 | | | 7 | 8 | | | | | |
| 5 | | | | 9 | 10 | | | | |
| 6 | | | | | 11 | 12 | | | |
| 7 | | | | | | 13 | 14 | | |
| 8 | | | | | | | 15 | 16 | |
| 9 | | | | | | | | 17 | 18 |

## Subtraction Strip Board

| 1 | 2 | 3 | 4 | 5 | 6 | 7 | 8 | 9 | 10 | 11 | 12 | 13 | 14 | 15 | 16 | 17 | 18 |
|---|---|---|---|---|---|---|---|---|---|---|---|---|---|---|---|---|---|
| | | | | | | | | | | | | | | | | | |
| | | | | | | | | | | | | | | | | | |
| | | | | | | | | | | | | | | | | | |
| | | | | | | | | | | | | | | | | | |
| | | | | | | | | | | | | | | | | | |
| | | | | | | | | | | | | | | | | | |
| | | | | | | | | | | | | | | | | | |
| | | | | | | | | | | | | | | | | | |
| | | | | | | | | | | | | | | | | | |
| | | | | | | | | | | | | | | | | | |
| | | | | | | | | | | | | | | | | | |
| | | | | | | | | | | | | | | | | | |

# Subtraction Control Chart 1

| | | | | | | | | | | | | | | | | | |
|---|---|---|---|---|---|---|---|---|---|---|---|---|---|---|---|---|---|
| 1 - 1 = 0 | 2 - 2 = 0 | 3 - 3 = 0 | 4 - 4 = 0 | 5 - 5 = 0 | 6 - 6 = 0 | 7 - 7 = 0 | 8 - 8 = 0 | 9 - 9 = 0 | 10 - 9 = 1 | 11 - 9 = 2 | 12 - 9 = 3 | 13 - 9 = 4 | 14 - 9 = 5 | 15 - 9 = 6 | 16 - 9 = 7 | 17 - 9 = 8 | 18 - 9 = 9 |
| | 2 - 1 = 1 | 3 - 2 = 1 | 4 - 3 = 1 | 5 - 4 = 1 | 6 - 5 = 1 | 7 - 6 = 1 | 8 - 7 = 1 | 9 - 8 = 1 | 10 - 8 = 2 | 11 - 8 = 3 | 12 - 8 = 4 | 13 - 8 = 5 | 14 - 8 = 6 | 15 - 8 = 7 | 16 - 8 = 8 | 17 - 8 = 9 | |
| | | 3 - 1 = 2 | 4 - 2 = 2 | 5 - 3 = 2 | 6 - 4 = 2 | 7 - 5 = 2 | 8 - 6 = 2 | 9 - 7 = 2 | 10 - 7 = 3 | 11 - 7 = 4 | 12 - 7 = 5 | 13 - 7 = 6 | 14 - 7 = 7 | 15 - 7 = 8 | 16 - 7 = 9 | | |
| | | | 4 - 1 = 3 | 5 - 2 = 3 | 6 - 3 = 3 | 7 - 4 = 3 | 8 - 5 = 3 | 9 - 6 = 3 | 10 - 6 = 4 | 11 - 6 = 5 | 12 - 6 = 6 | 13 - 6 = 7 | 14 - 6 = 8 | 15 - 6 = 9 | | | |
| | | | | 5 - 1 = 4 | 6 - 2 = 4 | 7 - 3 = 4 | 8 - 4 = 4 | 9 - 5 = 4 | 10 - 5 = 5 | 11 - 5 = 6 | 12 - 5 = 7 | 13 - 5 = 8 | 14 - 5 = 9 | | | | |
| | | | | | 6 - 1 = 5 | 7 - 2 = 5 | 8 - 3 = 5 | 9 - 4 = 5 | 10 - 4 = 6 | 11 - 4 = 7 | 12 - 4 = 8 | 13 - 4 = 9 | | | | | |
| | | | | | | 7 - 1 = 6 | 8 - 2 = 6 | 9 - 3 = 6 | 10 - 3 = 7 | 11 - 3 = 8 | 12 - 3 = 9 | | | | | | |
| | | | | | | | 8 - 1 = 7 | 9 - 2 = 7 | 10 - 2 = 8 | 11 - 2 = 9 | | | | | | | |
| | | | | | | | | 9 - 1 = 8 | 10 - 1 = 9 | | | | | | | | |

# Stamp Game

The Stamp Game will be used extensively during this phase of your child's math journey. You can use the provided printables to create your own Stamp Game. (See Directions below.) Before you prepare your printable Stamp Game, please know that there are physical Montessori Stamp Games available on Amazon and other Montessori shops for around $25. They are made of quality wood and come with a wooden storage tray and lid.

**Directions for Printables:**

1. Print the stamps, skittles, and circle discs on quality card stock paper. (If you purchased a printed version of this workbook, the print quality is high enough to not need further lamination. Go to montessoriforhomeschoolers.com/pages/math-book-3 if you need to print any extra pages for lost stamps!)
2. Cut out the individual stamps, skittles, and discs.
3. Store each type of stamp in its own zip-top clear plastic bag or other storage organizer.
4. Store the skittles together in a single bag.
5. Store the discs together in a single bag.
6. Remove the organizer from the workbook and store the organizer, and the baggies of stamps, skittles, and discs in a folder alongside the Workbook.
7. When your child is working with the Stamp Game, first empty the plastic bags onto their proper sections of the paper organizer.

**Additional Ideas for Printable Stamp Game:**

1. ***Decoupage:*** Use a watered-down glue to decoupage the cut stamps onto wooden squares.
2. ***Foam Board:*** Glue the sheets to a thin sheet of foam, then cut with a utility knife.
3. ***Laminate***: You may prefer to laminate your stamps, but I find that the corners are quite sharp. Rounding that many corners is a tedious job, but if you are up for it, do it!
4. ***Make your own:*** Find some one-inch squares, red, blue, and green paint, some sealer, and a white paint pen, and you can have a lovely DIY Stamp Game!
5. ***Game Pieces and Counters:*** You can use game pieces and counters for the skittles and discs.

# My Stamp Game

| | | | | | | |
|---|---|---|---|---|---|---|
| 1000 | 1000 | 1000 | 1000 | 1000 | 1000 | 1000 |
| 1000 | 1000 | 1000 | 1000 | 1000 | 1000 | 1000 |
| 1000 | 1000 | 1000 | 1000 | 1000 | 1000 | 1000 |
| 1000 | 1000 | 1000 | 1000 | 1000 | 1000 | 1000 |
| 1000 | 1000 | 1000 | 1000 | 1000 | 1000 | 1000 |
| 1000 | 1000 | 1000 | 1000 | 1000 | 1000 | 1000 |
| 1000 | 1000 | 1000 | 1000 | 1000 | 1000 | 1000 |
| 1000 | 1000 | 1000 | 1000 | 1000 | 1000 | 1000 |
| 1000 | 1000 | 1000 | 1000 | 1000 | 1000 | 1000 |

| | | | | | | |
|---|---|---|---|---|---|---|
| 100 | 100 | 100 | 100 | 100 | 100 | 100 |
| 100 | 100 | 100 | 100 | 100 | 100 | 100 |
| 100 | 100 | 100 | 100 | 100 | 100 | 100 |
| 100 | 100 | 100 | 100 | 100 | 100 | 100 |
| 100 | 100 | 100 | 100 | 100 | 100 | 100 |
| 100 | 100 | 100 | 100 | 100 | 100 | 100 |
| 100 | 100 | 100 | 100 | 100 | 100 | 100 |
| 100 | 100 | 100 | 100 | 100 | 100 | 100 |
| 100 | 100 | 100 | 100 | 100 | 100 | 100 |

| | | | | | | |
|---|---|---|---|---|---|---|
| 10 | 10 | 10 | 10 | 10 | 10 | 10 |
| 10 | 10 | 10 | 10 | 10 | 10 | 10 |
| 10 | 10 | 10 | 10 | 10 | 10 | 10 |
| 10 | 10 | 10 | 10 | 10 | 10 | 10 |
| 10 | 10 | 10 | 10 | 10 | 10 | 10 |
| 10 | 10 | 10 | 10 | 10 | 10 | 10 |
| 10 | 10 | 10 | 10 | 10 | 10 | 10 |
| 10 | 10 | 10 | 10 | 10 | 10 | 10 |
| 10 | 10 | 10 | 10 | 10 | 10 | 10 |

| | | | | | | |
|---|---|---|---|---|---|---|
| 1 | 1 | 1 | 1 | 1 | 1 | 1 |
| 1 | 1 | 1 | 1 | 1 | 1 | 1 |
| 1 | 1 | 1 | 1 | 1 | 1 | 1 |
| 1 | 1 | 1 | 1 | 1 | 1 | 1 |
| 1 | 1 | 1 | 1 | 1 | 1 | 1 |
| 1 | 1 | 1 | 1 | 1 | 1 | 1 |
| 1 | 1 | 1 | 1 | 1 | 1 | 1 |
| 1 | 1 | 1 | 1 | 1 | 1 | 1 |
| 1 | 1 | 1 | 1 | 1 | 1 | 1 |

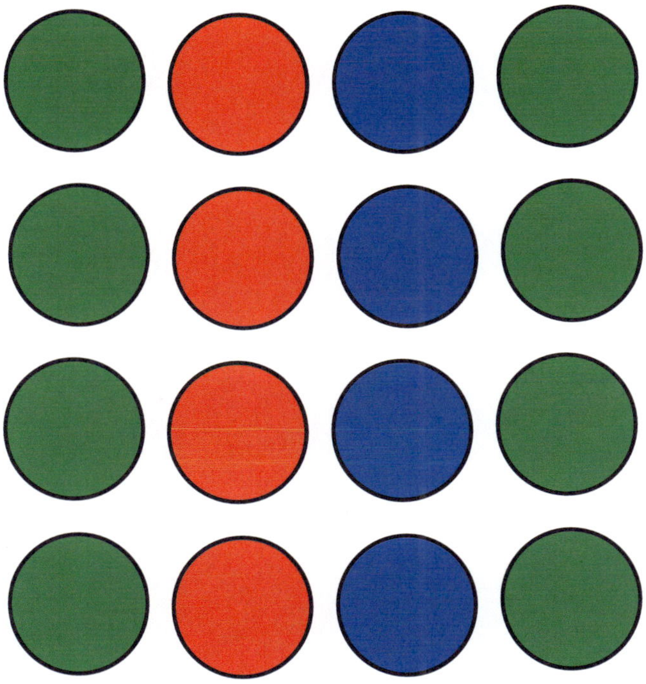

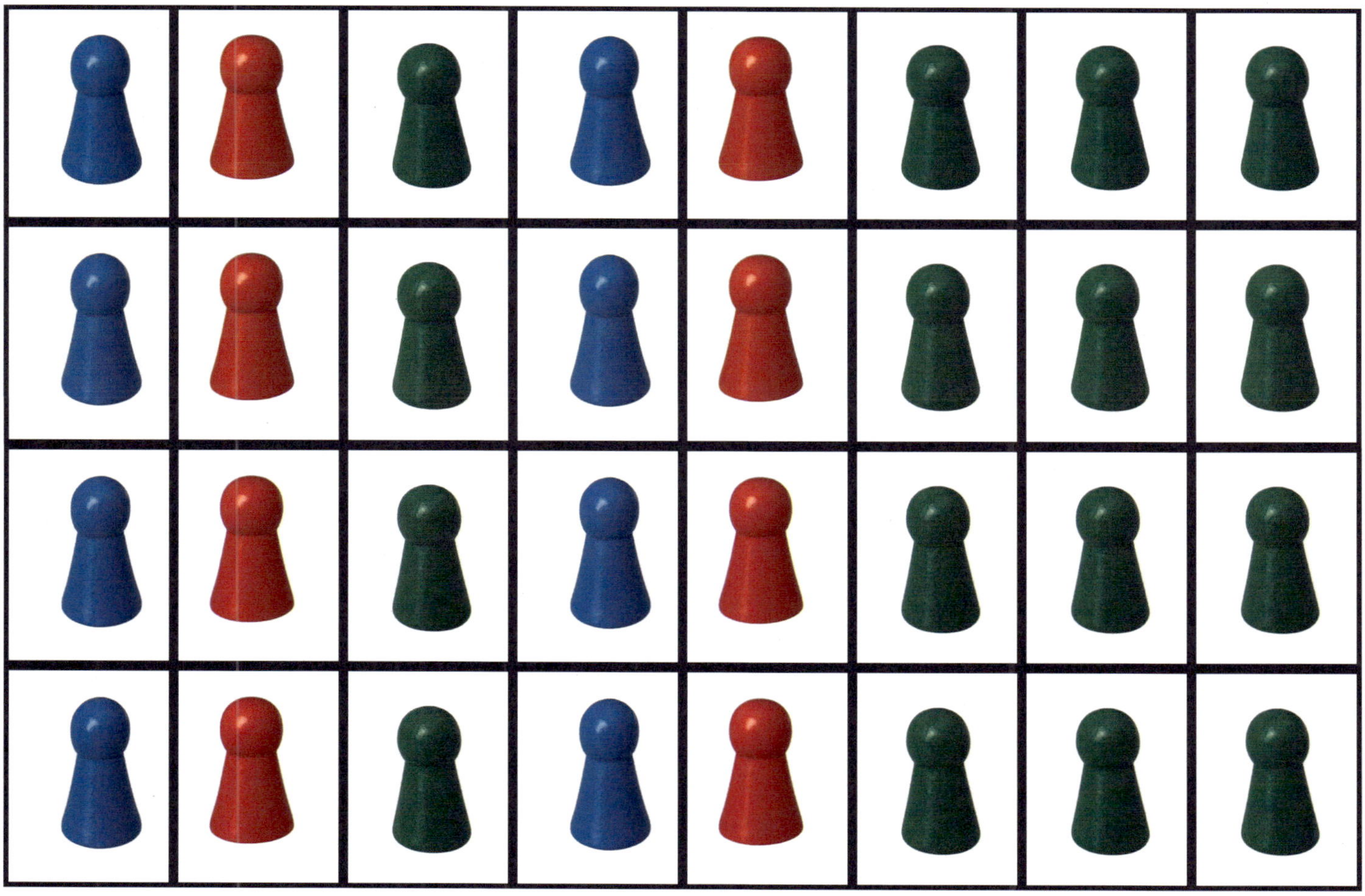

# Stamp Game Solutions



## 6.2 | Stamp Game: Addition Problems

1. 1573 + 6214 = 7787
2. 2475 + 827 = 3302
3. 8022 + 708 = 8730
4. 6352 + 3251 = 9603
5. 206 + 822 = 1028
6. 58 + 94 = 152

83

7. 2571 + 279 = 2850
8. 1298 + 584 = 1882
9. 4353 + 568 = 4921
10. 387 + 284 = 671
11. 91 + 44 = 135
12. 78 + 65 = 143

84

## 7.2 | Stamp Game: Subtraction Problems

1. 1677 − 214 = 1463
2. 3802 − 965 = 2837
3. 4123 − 2383 = 1740
4. 1413 − 462 = 951
5. 138 − 85 = 53
6. 4677 − 1394 = 3283

89

7. 960 − 147 = 813
8. 506 − 185 = 321
9. 8038 − 2926 = 5112
10. 3221 − 675 = 2546
11. 86 − 37 = 49
12. 479 − 181 = 298

90

## 8.2 | Stamp Game: Multiplication Problems

| # | Problem | Answer |
|---|---|---|
| 1 | 2133 × 3 | 6399 |
| 2 | 2215 × 3 | 6645 |
| 3 | 4322 × 2 | 8644 |
| 4 | 2243 × 3 | 6729 |
| 5 | 3610 × 2 | 7220 |
| 6 | 1524 × 3 | 4572 |

95

| # | Problem | Answer |
|---|---|---|
| 7 | 2445 × 3 | 7335 |
| 8 | 2344 × 3 | 7032 |
| 9 | 4752 × 2 | 9504 |
| 10 | 2148 × 3 | 6444 |
| 11 | 3010 × 2 | 6020 |
| 12 | 524 × 3 | 1572 |

96

## 8.5 | Stamp Game: Multiplication with 2-Digit Multiplier

**Directions:**
Complete the problems using the green unit skittles. For example, in problem #1, place 11 unit skittles in a column and lay out stamps for the number 231 next to each green skittle. Add them all together at the end, making exchanges as necessary.

1. 11 × 231 = 2541
2. 12 × 124 = 1488
3. 10 × 311 = 3110
4. 13 × 242 = 3146

99

## 9.2 | Stamp Game: Division Problems

| # | Problem | Answer |
|---|---|---|
| 1 | 9633 ÷ 3 | 3211 |
| 2 | 5644 ÷ 4 | 1411 |
| 3 | 4928 ÷ 4 | 1232 |
| 4 | 9642 ÷ 3 | 3214 |
| 5 | 5560 ÷ 5 | 1112 |
| 6 | 7842 ÷ 3 | 2614 |

106

## 9.3 | Stamp Game: Division Problems with Remainders

Sometimes when we divide, we have stamps leftover that cannot be shared equally among the skittles. We call the leftover stamps the "remainder." Each of the problems on this page has a remainder. Write the quotient, then write the remainder in the box.

1. 4873 ÷ 4 = 1218 R = 1
2. 6815 ÷ 3 = 2271 R = 2
3. 4325 ÷ 2 = 2162 R = 1
4. 7558 ÷ 5 = 1311 R = 3
5. 7869 ÷ 6 = 1311 R = 3
6. 5147 ÷ 3 = 1715 R = 2

107

## 9.5 | Stamp Game: Division with 2-Digit Divisor Problems (no exchanges)

1. 2553 ÷ 23 = 111
2. 6752 ÷ 32 = 211
3. 3926 ÷ 13 = 302
4. 2814 ÷ 14 = 201
5. 1919 ÷ 19 = 101
6. 4662 ÷ 42 = 111

111

## 9.7 | Stamp Game: Division with 2-Digit Divisor - Advanced Problems

**Directions**:
Follow the directions leading up to this problem. You will need to make exchanges. Remember to use one blue skittle to represent each group of "ten people" and one green skittles to represent every "one person."

1. 2752 ÷ 16 = 172
2. 2628 ÷ 18 = 146
3. 3528 ÷ 14 = 252
4. 6372 ÷ 27 = 236
5. 9548 ÷ 44 = 217
6. 7956 ÷ 34 = 234

118

## 9.8 | Stamp Game: Division with Zero in Dividend

**Directions**:
1. Perform the division in the same manner as 2-digit division.
2. In order to get the necessary stamps under each skittle, you will need to make an exchange.
3. Write any remainders on the line. If there is no remainder, you can leave it blank or write "0."

1. 1308 ÷ 12 = 109 R = 0
2. 2850 ÷ 15 = 190 R = 0
3. 3303 ÷ 22 = 150 R = 3
4. 4065 ÷ 26 = 156 R = 9

119

## 9.9 | Stamp Game: Division with Zero in Divisor

**Directions**:

1. Perform the division in the same manner as 2-digit division. You will use the red, blue, and green skittles or discs. If there is a *zero* in the divisor, then you need to use a *disc* for that category. No stamps will be given to that category.
2. In order to get the necessary stamps under each skittle, you may need to make an exchange.

① 2244 ÷ 102 = 22

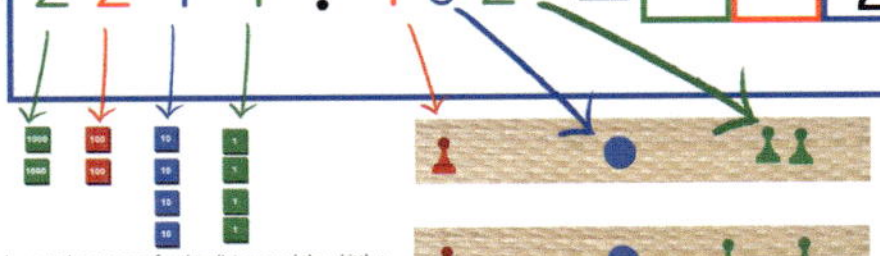

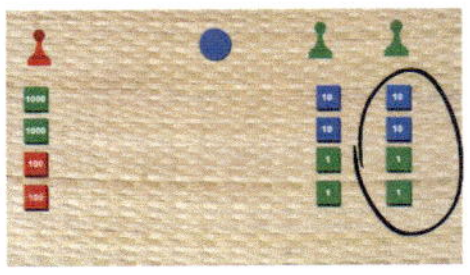

1. First, lay out the stamps for the divisor and the skittles and disc for the dividend.
2. Next, distribute the thousands, hundreds, and tens, going down in stamp value for each category in the divisor.
3. Wait! There are no "people" in the hundreds category, so do not give that category any stamps. In this problem, that means you won't distribute the hundreds yet.
4. Now distribute the hundreds and units.
5. Last, distribute the units.
6. Write down your answer.

② 8844 ÷ 402 = 22

③ 5992 ÷ 107 = 56

④ 5427 ÷ 603 = 9

120

② 9791 ÷ 110 = 89 R = 1

③ 6155 ÷ 150 = 41 R = 5

④ 3680 ÷ 160 = 23 R = 0

122

Book 3 Completed

Name:________________________

Date Completed:____________

I can memorize addition and subtraction facts, work with the Stamp Game, and tell time!

Dear ________________,

You did it! You worked hard and completed your **Montessori Math Workbook, Book 3!** I am most proud of you for ____________________________.
I love how you ____________________________
________________________________________.
You have proven you are capable of
________________________________________!
Most of all, I want you to know that *I love you*, and we are on this journey together, every step of the way.

Love, ____________________

Made in United States
Cleveland, OH
05 November 2026

47637960R00148